SOLID STATE PHYSICS-1

Institute of Fundamental Studies
Kandy, Sri Lanka

SOLID STATE PHYSICS-1

(Proceedings of the First International Symposium on Solid State Physics held in Sri Lanka on April 20-25, 1987)

Edited by M.A.K.L. Dissanayake,
R. Attele and K. Tennakone

Nova Science Publishers
Commack•Budapest

Nova Science Publishers, Inc.
283 Commack Road
Suite 300
Commack, New York 11725

Library of Congress Cataloging-in-Publication Data available upon request

ISBN 0-941743-66-7

Graphic Design by Elenor Kallberg and Peggy Harvey

Printed in the United States of America

TABLE OF CONTENTS

INTRODUCTION

Although a number of research groups in the Asian region are actively working on different areas of solid state research, opportunities of them to meet and interact among themselves and with each other are rather limited. The primary objective of the First International Symposium on Solid State Physics, held at the Institute of Fundamental Studies was to provide a forum for researchers in the region to meet and exchange ideas and to expose young scientists to new and exciting developments in Solid State Physics. The symposium was attended by eleven invited speakers and thirty-one participants from five countries. The program consisted of overview lectures given by invited speakers and short presentations by participants, covering the areas of high temperature superconductivity, semiconductors, and solid electrolytes. This volume contains the overview lectures and the short presentations.

The success of the symposium was primarily due to the encouragement given by Professor Cyril Ponnamperuma, the Director of the Institute of Fundamental Studies. Organization of this symposium would not have been possible without support from the International Centre for Theoretical Physics and the Third World Academy of Sciences, Trieste, Italy. Financial support was also obtained from the British Council, Colombo, Sri Lanka, and the International Seminar on Physics, Uppsala University, Sweden. Our deep appreciation goes to Professor Abdus Salam, Director ICTP who took a personal interest in making this symposium a success.

PREFACE

It is with great pleasure that I welcome the participation of the proceedings of the first major academic gathering sponsored by the Kandy School of Science (KSS). Indeed, the primary objective of the KSS has been to act as an international forum for science in the North/South, and South/South dialogue. It is perhaps not inappropriate to mention that the very concept of the School arise in a conversation between myself and Nobel Laureate Abdus Salam during his visit to Sri Lanka in late 1986. The KSS has therefore been organized from its inception by the Institute of Fundamental Studies in collaboration with the Third World Academy of Sciences and the International Centre for Theoretical Physics. It will provide a center for meetings, discussions, workshops, and training programs. We hope that the KSS will host seminars on subjects at the very frontiers of science. I am glad to say that this hope was realized in the very first seminar, the proceedings of which are made available to you now.

I am sure that this will be the first of a series of stimulating seminars and meetings sponsored by the KSS.

March 1988
Institute of Fundamental Studies
Kandy, Sri Lanka

Cyril Ponnamperuma
Director

MESSAGE FROM PROFESSOR ABDUS SALAM

Director, Centre for Theoretical Physics, and
President, Third World Academy of Sciences, Trieste, Italy

I am delighted that the first meeting of the Kandy School of Science has materialized. As you know, I fell in love with Kandy, with its atmosphere of scholarship, learning, and serenity, the very first time I was there. I am sure the School will be a great success. Best regards.

Our thanks go to all the invited lecturers and participants who attended the symposium and took a keen interest in the lectures and discussions and to the IFS staff involved in the organization of the symposium. Thanks are due to Mrs. Phyllis De Alwis for typing the manuscript.

March 15, 1988
Institute of Fundamental Studies
Hantana Road, Kandy

M. A. K. L. Dissanayake
R. Attele
K. Tennakone
Editors

PART 1

OVERVIEW LECTURES

QUANTUM PHENOMENON IN SMALL SEMICONDUCTOR STRUCTURES AND DEVICES

K. F. Berggren
Department of Physics and Measurement Technology
Linköping University
S-581 83 Linköing
Sweden

Abstract: **The ability to artificially structure new semiconductor materials on an atomic scale, using advanced crystal growth methods such as molecular beam epitaxy and metal organic chemical vapor deposition, has recently led to the observation of new physical phenomena as well as to the creation of entirely new classes of devices based on bandgap and wave function engineering. In this lectures an elementary introduction is given to the quantum aspects of these new structures.**

INTRODUCTION

New material growth technologies have made it possible to fabricate low-dimensional structures and structures with extremely small spatial extensions. Usual methods involve molecular beam epitaxial growth (MBE) and metal organic chemical vapor deposition (MOCVD) combined with (electron) lithography. In such structures the charge carriers do not behave in the same way as in bulk material and it is possible to construct devices with improved performance as well as new devices which make direct use of the properties of very small structures.

In this review we will give a brief and elementary introduction to selected quantum phenomena in some small semiconductor structures. The field is growing very rapidly and it is impossible to give complete up-to-date coverage. The number of research papers and conference proceedings is ever increasing, but the reader should find Refs. [1-5] useful for further reading. Also, the number of conferences, workshops, and schools is increasing at a rate that could keep the ambitious researcher almost constantly traveling around the world. Recent announcements for events during 1987 tell about "Physics and Applications of Quantum Wells and Superlattices", Erice, Sicili, Italy; "Third International Conference on Modulated Semiconductor Structures", Montpellier, France; "EP2DS7-Electronic Properties of Two-Dimensional Systems", Santa Fe, New Mexico, USA; "Meeting on Excitons in Confined Systems", Rome, Italy; "Bandgap Engineering: Heterojunctions, Superlattices, and Their Device Applications" and "Quantum Well and Superlattice Physics", Bay Point, Florida, USA; "Interfaces, Quantum Wells and Superlattices", Banff, Alberta, Canada; "Fourteenth International Symposium on GaAs and Related Compounds", Heraklio, Crete, Greece; "Superlattices, Microstructures and Microdevices", Chicago, USA, etc. Most major meetings on condensed matter physics and materials science are also likely to have special sessions on small structures and nanophysics.

Although we should really talk about the symbiosis of technology and basic research, we may oversimplify here somewhat by saying that one distinguishes two mainstreams in the present development. One is more technological and is oriented towards design of new devices of immediate practical interest. Such R&D is primarily conducted at industrial laboratories. The other mainstream, predominantly found at more academic institutions, concerns condensed matter physics itself. The reason for this is that low-dimensional semiconductor structures, for example, offer novel and unique model systems that have led to a renewal of fundamental condensed matter physics. The integer quantized Hall effect (QHE) [6] is probably the most spectacular discovery of this kind and Hall was awarded the Nobel prize in physics 1985. In brief, QHE says that ρ_{xy}, the xy-component of the resistivity of a two-dimensional electrongas in a perpendicular field, assumes quantized values h/ie^2 (integer i, Planck's constant h) when the chemical potential lies in the regime of localized states between Landau levels. The accuracy of this relation is better than 10^{-7}. The fractional quantum Hall effect (FQHE) [6] is equally fascinating. Hall plateaus are seen at quantized resistivity values $h/\nu e^2$ where ν = 1/3, 1/5, 2/5, 2/7, 3/7, 4/9, etc. correspond to a fractional filling of the lowest Landau state. Experimentally the two effects look very similar, but the underlying physics is quite different. The FQHE is thought to be due to the formation of a new state of matter, the incompressible quantum fluid, in which a two-dimensional electron gas of interacting electrons has condensed into a many-body ground state. The IQHE, on the other hand, is discussed in terms of one-electron states and localization due to disorder. Modern semiconductor technology has also opened up other areas of fundamental research: subband spectroscopy [1], localization in low-dimensional systems due to disorder and electron interactions, metal-insulator transition, universal quantum fluctuations [3], quantum size effects, quasiperiodic Fibonacci superlattices [7], and the Aharonov-Bohm effect [8] to mention a few.

The following introduction to a vast and quickly expanding field must by necessity be limited to only a few subjects. The list of references should, however, be useful in tracing the current literature. In Section 2 we describe some common simplifications of electron states in semiconductors. The quantum well concept and subbands are introduced in Section 3, and the realization of such wells is described in Section 4. Quantum transport is discussed in Section 5 and the effect of an external magnetic field is dealt with in Section 6. Systems other than semiconductor structures are briefly mentioned in Section 7. Some concluding remarks are given in Section 8.

2. Electron States in Semiconductors and Some Simple Cations

Before going to a more detailed discussion of electron states in various devices it is appropriate to specify our theoretical framework. In an ordinary crystalline system the one-electron wave functions are the Bloch functions

$$\psi_{n,\vec{k}}(\vec{r}) = e^{i\vec{k}\cdot\vec{r}} u_{n,\vec{k}}(\vec{r}) \tag{2.1}$$

where n is the band index, $\vec{k}$ is the crystal wave vector, and $u_{n,\vec{k}}(\vec{r})$ is a function with the same periodicity as the lattice. The wave function satisfies the Schrödinger equation

$$\left[-\frac{\hbar^2}{2m_e}\Delta + V(\vec{r})\right] \psi_{n,\vec{k}}(\vec{r}) = \varepsilon_n(\vec{k})\, \psi_{n,\vec{k}}(\vec{r}) \tag{2.2}$$

where m_e is the bare electron mass, $V(\vec{r})$ is an effective one-electron potential and $\varepsilon_n(\vec{k})$ is the eigenenergy defining the bandstructure of the material. Solving Equation (2.2) is generally a complex numerical problem. Since electronic processes in semiconductors and semiconductor devices usually only involve charge carriers associated with states close to the bottom of conduction bands (electrons) or to the top of the valence bands (holes) there is less need to have a complete knowledge of the bandstructure and corresponding Bloch functions. Rather, we may resort to the effective mass theory [9] (EMT), which often brings the problem to the level of elementary textbook quantum mechanics. Consider, for example, an isotropic conduction band minimum. For states with energies in the vicinity of this minimum the full Schrödinger equation in (2.2) is then replaced by the effective equation

$$\left[-\frac{\hbar^2}{2m_c^*}\Delta + V(\vec{r})\right] \varphi(\vec{r}) = E\varphi(\vec{r}) \tag{2.3}$$

where m_c^* is the conduction band effective mass and $V_e(\vec{r})$ is a potential term describing the difference between the real potential (including, for example, dopant) and the perfect crystal potential in Equation (2.2); $\varphi(\vec{r})$ is an envelope wave function, and E is the energy relative to the unperturbed conduction band minimum. This description is well known in connection with shallow hydrogenic donor states. Then $V_e(\vec{r})$ is simply the Coulomb potential reduced by the macroscopic dielectric constant κ of the host material, $\varphi(\vec{r})$ is a hydrogenic solution scaled by an effective Bohr radius, and E is the sequence of scaled hydrogenic levels.

Similarly an electron injected near the bottom of the conduction band is simply described by a plane wave. A heavily doped semiconductor displaying metallic conduction at low temperatures may be viewed as an electron (hole) gas occupying the unperturbed conduction (valence) band and moving in the background of ionized impurities, dielectrically screened by the electron (hole) gas (see Figure 1). In this sense the description of a doped semiconductor becomes exceedingly simple and we may view a doped semiconductor as an ideal model system for studying the properties of the electron (hole) gas. This elementary free gas model is also a realistic starting point for incorporating interaction effects and impurity scattering by perturbative in methods in order to study bandgap modifications due to heavy doping [10, 11]. Such effects are becoming increasingly important to

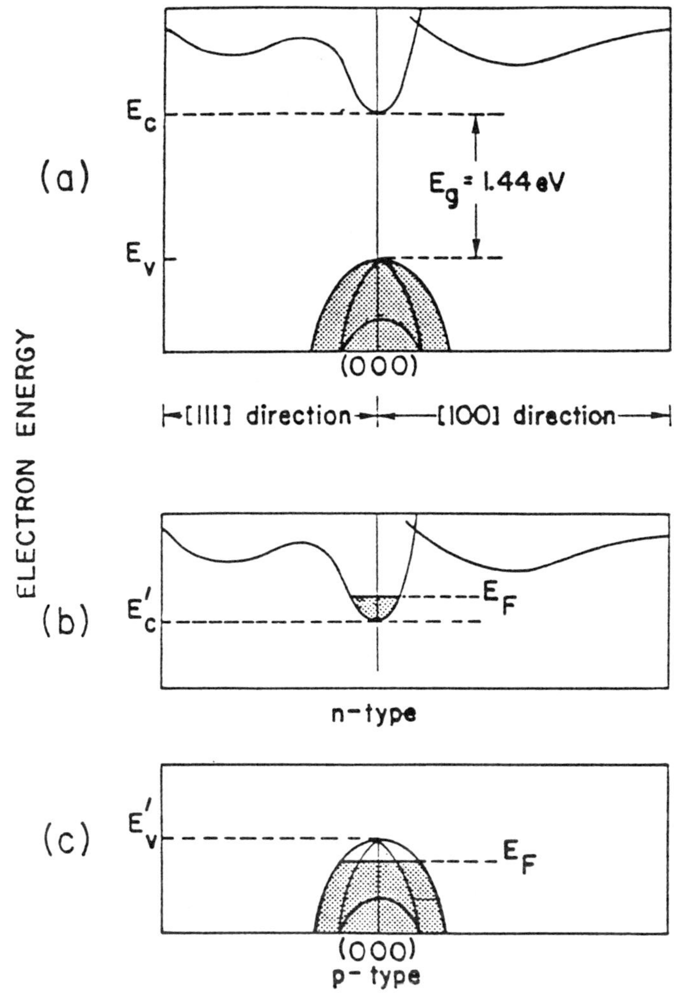

Fig. 1. (a) shows the bands just above and below the intrinsic gap for GaAs. In a heavily doped *n*-type material (b) doped in excess of Mott's critical concentration for the metal-non-metal transition electrons are assumed to occupy states near the bottom of the conduction band in the form of a free electron gas; (c) illustrates the case of a hole gas associated with *p*-doping. Because of interaction effects and scattering off ionized impurities bands are slightly shifted (bandgap modification, $E_C \neq E^{a}$, $E_V \neq E_V'$).

large integrated circuits as geometries shrink and doping levels necessarily increase.

The effective mass theory is attractive from many points of view. It is simple and for many purposes also quantitatively quite reliable. When discussing electron (hole) states in various devices in the following sections we shall have EMT in mind. Although this is a standard approach we must not forget that EMT is an approximate theory and that there are limits to its applicability.

3. The Quantum Well Concept and Subbands

Design of semiconductor devices usually embraces two different aspects. The first is material theory; this branch of physics relies of quantum mechanics as a powerful tool for the understanding of basic material properties. These material parameters are then used as given quantities in the second part of the design. In this step the device behavior is described by concepts which involve the basic material parameters, but do not use quantum mechanics directly. These two aspects of design are often distinct, and a device design can be performed quite adequately with only a limited or no appreciation of quantum mechanics.

Quantum mechanics deals with waves. This wave aspect becomes important when the corresponding wavelengths are comparable to some characteristic device length. Since refined growth methods for semiconductor crystals have now made it possible to fabricate structures with dimensions comparable to the wavelength ascribed to the charge carriers, quantum mechanical effects are directly influencing device design. In other words, the two aspects of design mentioned are no longer decoupled.

In a semiconductor structure consisting of a very thin layer of semiconducting material, sandwiched between a winder bandgap material, we can have a situation in which charge carriers are trapped in a potenti8al well. The behavior of an electron in such a well is one of the standard textbook examples in quantum mechanics. This example is easy to handle and reveals some of the important consequences of confinement. Here we consider an infinitely deep square well (Figure 2). If t is the width of the well, the normalized solutions of the stationary states are:

$$\varphi_n(x) = \sqrt{2/t}\,\sin(\pi n x/t), \tag{3.1}$$

where n is a positive integer. The corresponding energy levels are

$$E_n = \frac{\hbar^2}{2m^*}\left(\frac{n\pi}{t}\right)^2 \tag{3.2}$$

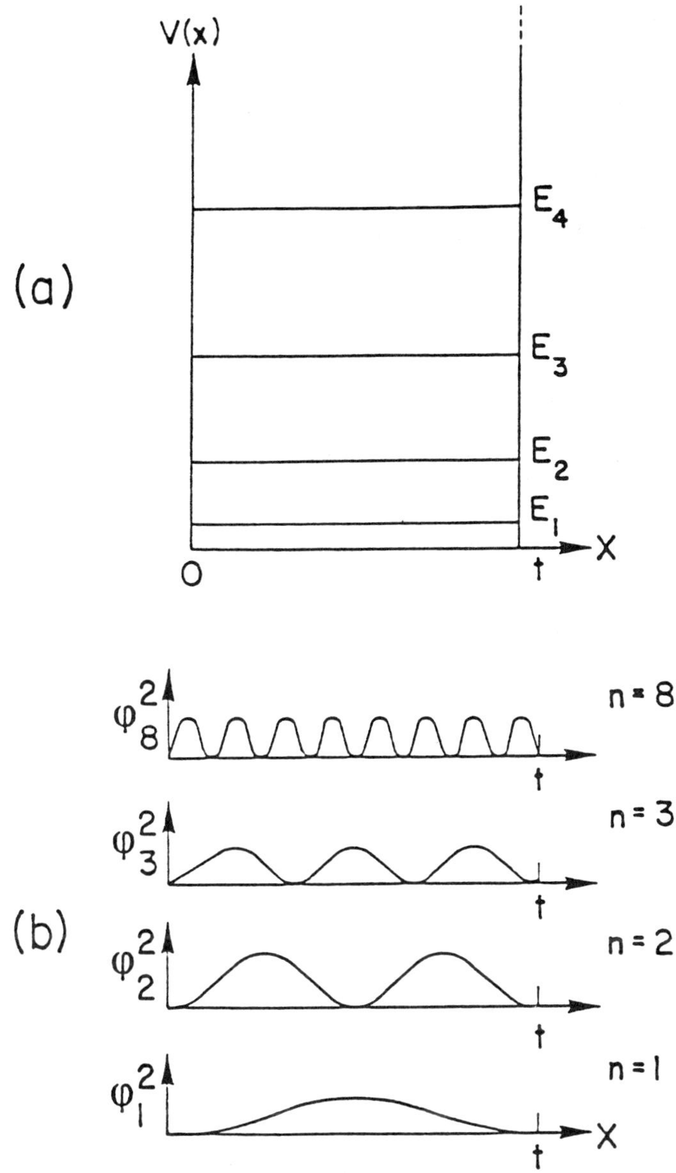

Fig. 2. (a) shows the potential V as a function of x for an infinite square well, along with energy levels of the four lowest energy states. The probability distributions corresponding to energy states with n = 1, 2, 3, and 8 are shown in (b)

where m^* is the effective mass of the semiconductor. Because of the confinement the transverse motion is thus geometrically quantized and a discrete set of sublevels is formed (Figure 2). This is usually referred to as a quantum size effect (QSE). The motion parallel to the layer remains free. The complete, normalized wave function of an electron is therefore:

$$\varphi_{n,\vec{k}}(\vec{r}) = \varphi_n(x)^{ik_{\parallel} \cdot r_{\parallel}}/\sqrt{A} \tag{3.3}$$

where A is the cross-sectional area and $k_{\parallel}$ denotes the parallel wave vector. The corresponding energies

$$E_n(k_{\parallel}) = E_n + \frac{\hbar^2 k_{\parallel}^2}{2m^*} \tag{3.4}$$

define two-dimensional (2D) *subbands*. The density of states per unit area, $g_{2D}^n(E)$, associated with such a 2D subband is a constant,

$$g_2^n(E) = m^*/(\pi\hbar^2) \tag{3.5}$$

for $E \geqslant E_n$ and zero otherwise. The total density of states is the quasi two-dimensional form

$$g_{Q2D}(E) = m^*/(\pi\hbar^2)\sum_n \theta(E - E_n) \tag{3.6}$$

where $\theta(x) = 1$ for $x \geqslant 0$ and 0 if $x < 0$. Because of confinement the continuous three-dimensional (3D) density of states (per unit volume)

$$g_{3D}(E) = \frac{m^*}{\pi^2\hbar^3} \sqrt{2m^* E} \tag{3.7}$$

is thus broken into steps as illustrated in Figure 3. The steps increase in size with decreasing width of the layers.

As a somewhat more realistic example we consider a well of finite depth. Now there is only a finite number of discrete states which are different from those in an infinite well. For a very narrow, or for a very shallow well only one state is found. The wave functions of the bound states now decay exponentially into the regions outside the well. The considerations above can readily be extended to multiple quantum wells, superlattices, and wells other than rectangular wells. We will return to some of these cases in later sections.

In recent years, there has also been growing interest in the study of the properties of ultra-thin semiconductor wires, also called quantum well wires, with submicron dimensions. These are quasi one-dimensional (1D) structures in which the carriers are confined to move along the length of the

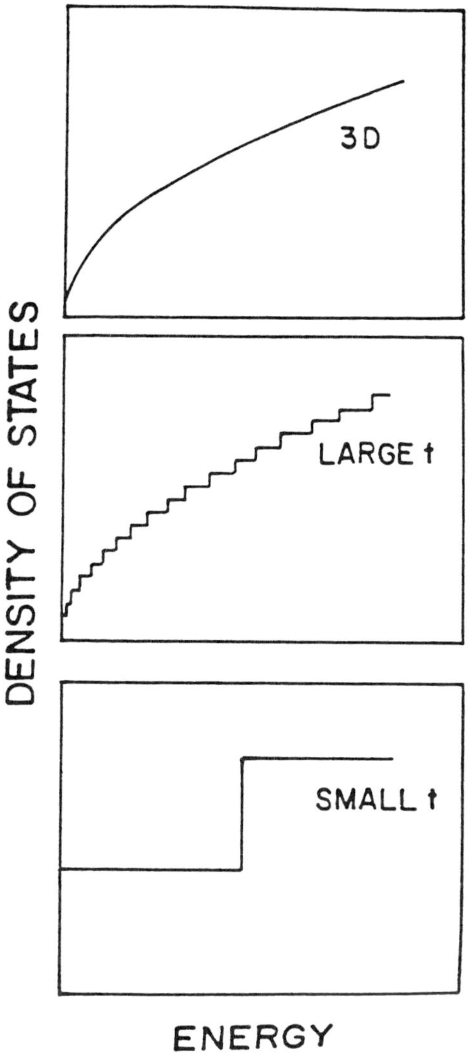

Fig. 3. Schematic behavior of the density of states associated with the electrons confined to a semiconductor layer or slab. if the layer is very wide the electrons form in practice a 3D electron gas with the usual continuous density of states $g(E) \alpha \sqrt{E}$. As the thickness t is decreased the 3D density of states breaks up into discrete subbands, each having a constant 2D density of states.

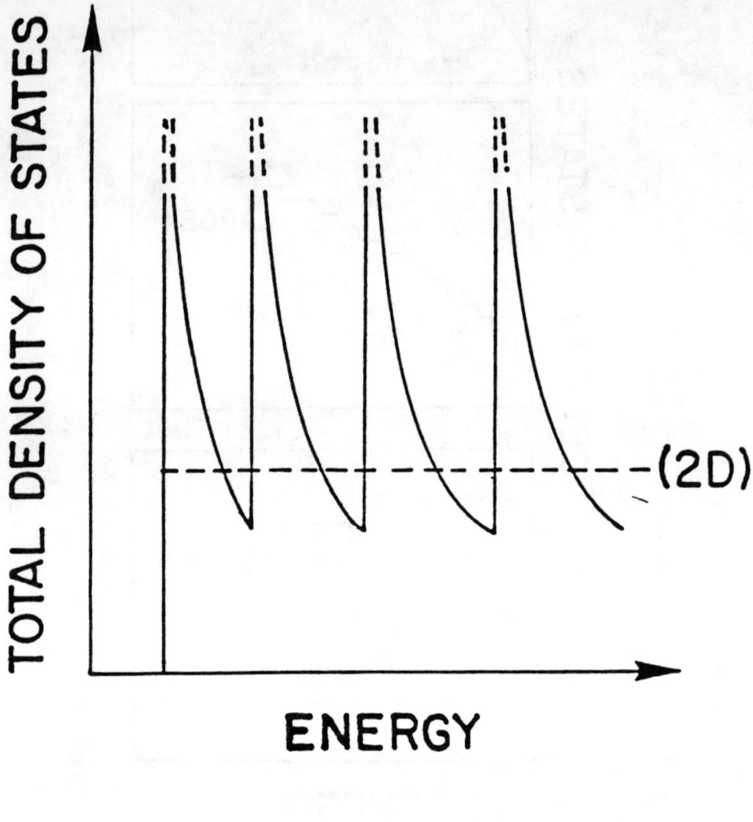

Fig. 4. A schematic view of the total density of states for a quasi 1D system. Each subband gives rise to a square-root singularity of equal strength.

wire but the motion is quantized in the transverse directions. Sakaki [12] was the first to investigate theoretically the impurity-limited mobility in such wires. The possibility of obtaining enhanced mobility led him to propose the fabrication of semiconductor wires, which could serve as promising tools for semiconductor devices like high speed quantum integrated circuits with low power consumption. Sakaki's model calculations have more recently been followed up by Kubakaddi et al. [13, 14]. To date an enhanced mobility has not been observed experimentally.

If the cross-section is taken to be rectangular with dimensions t_x and t_y the normalized one-electron states are

$$\varphi_{n1,k_z}(\vec{r}) = \frac{2}{\sqrt{t_x t_y L}} \sin(\pi n x/t_x) \sin(\pi l y/t_y) e^{ik_z z} \tag{3.8}$$

where n and l are positive integers and L is the length of the wire. The sublevels are

$$E_{n,l} = \frac{\hbar^2}{2m^*}\left[\left(\frac{n\pi}{t_x}\right)^2 + \left(\frac{l\pi}{t_y}\right)^2\right] \tag{3.9}$$

and the one-dimensional subbands

$$E_{nl}(k_z) = E_{n,l} + \frac{\hbar^2}{2m^*}k_z^2 \tag{3.10}$$

The corresponding density of states/unit length, $g_{1D}^{nl}(E)$, is the 1D form

$$g_{1D}^{nl}(E) = \frac{1}{\pi\hbar}\sqrt{2m^*/(E - E_{nl})}\;\theta(E - E_n) \tag{3.11}$$

The total quasi 1D density of states

$$g_{Q}lD(E) = \sum g_{1D}^{nl}(E) \tag{3.12}$$

is shown schematically in Figure 4 for the case $t_x \gg t_y$. The constant 2D density of states is split into sharply peaked 1D contributions. The picture somewhat resembles a 3D electron gas in a strong magnetic field. In a later section we will discuss the observation of 1D subbands in a particular quasi 1D structure by means of magnetic depopulation.

Although many of the efforts thus far have been directed towards two- and one-dimensional structures, the possibility of fabricating quasi-atomic systems with the use of a semiconductor heterostructure has been pointed out by Watanabe [15] on the basis of the current status and future prospect of microfabrication technology. These novel and highly speculative mesiscopic structures were named "superatoms" and "antisuperatoms" or "quantum balls." While the superatom may be useful from an applied point of view (use of a superatom array for functional circuits has been suggested [16]) we can also expect it to have unique physical properties both in itself and in the form of superatom "molecules" or "crystals." Some of these aspects have been pursued in Ref. [17].

4. Realization of Some 2D and 1D Quantum Wells

The bandgap and lattice constant of a semiconducting material will depend on the material composition. It is possible to fabricate semiconductors with different bandgaps, but with very similar lattice constants. At the junction between two such different semiconductors (a heterojunction) the bandgap differences will give rise to discontinuities, ΔE_v and ΔE_c, of the valence and conduction band edges. The line-up of the band edges depends on the particular materials. Different situations are illustrated schematically in Figure 5 but the details of the line-up are poorly understood in general. In the important case of the GaAlAs/GaAs heterojunction, the discontinuity is shared between the bands.

4.1. Rectangular Quantum Wells

For a structure like the one in Figure 6, the discontinuities of the bands form a potential well for both electrons and holes in the GaAs. If the thickness becomes less, on the order of a few hundred Angstroms or less, we start to have QSE. As described in the previous section, the momentum of the charge carriers perpendicular to the well becomes quantized and only discrete values for the energy corresponding to the momentum perpendicular to the well are allowed. The carriers can still move freely in the other two dimensions, and the corresponding energies are not restricted. The material therefore behaves as if the bandgap had been increased. Since the subband energies increase with decreasing thickness the exact value of this artificial bandgap can be tailored. This is of significant interest for device applications.

As mentioned in the previous section, the transverse wave functions extend a certain distance into the GaAlAs "barrier" regions. In a structure consisting of several wells separated by thin barriers (50 Å or less), that is, a superlattice, the wave function in adjacent wells will then overlap. If the overlap is strong the discrete levels of the single quantum well will broaden

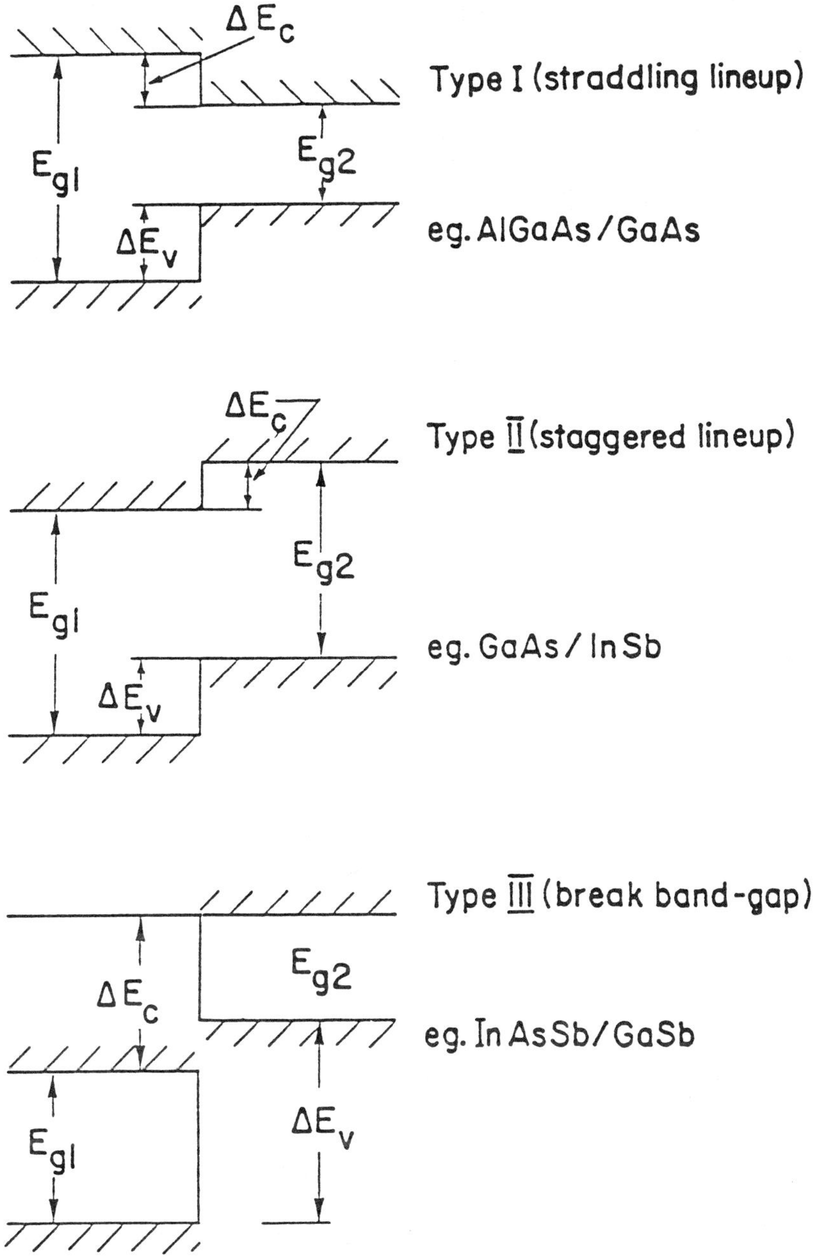

Fig. 5. Three types of band lineup at a heterojunction have been suggested.

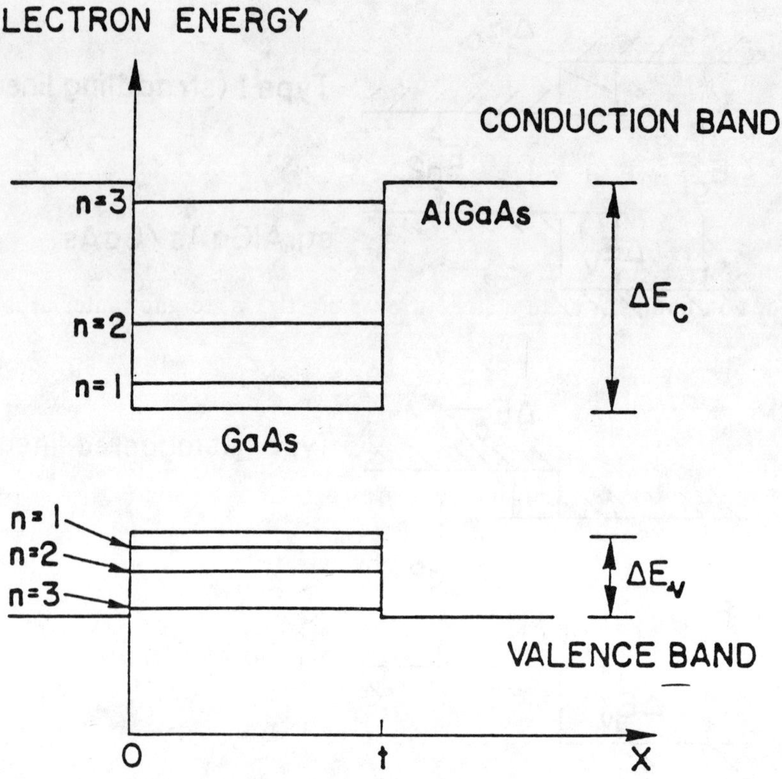

Fig. 6. Band edges of a quantum well in undoped GaAs/GaAlAs showing the first three confined energy levels. ΔE_c and ΔE_v are the bandgap discontinuities.

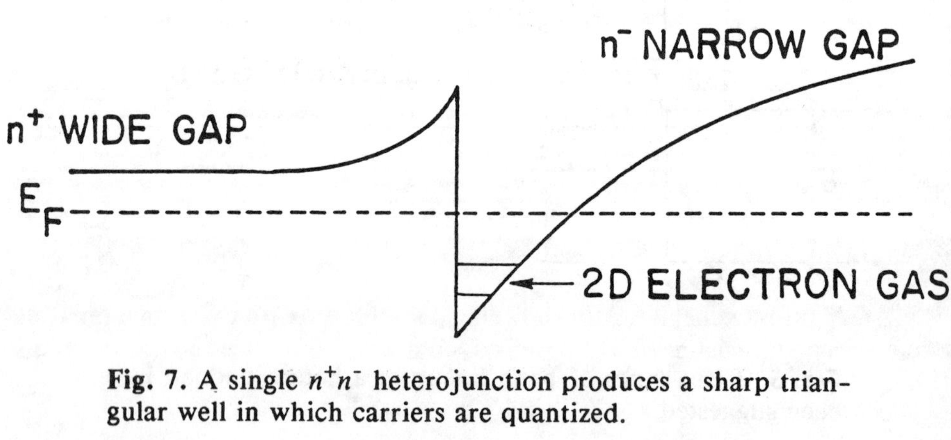

Fig. 7. A single n^+n^- heterojunction produces a sharp triangular well in which carriers are quantized.

into bands. Thus superlattices have properties different from both bulk material and single quantum wells.

4.2. Single Heterojunctions

If we look in detail at a single heterojunction between a wide and narrow gap semiconductor, a band bending can be so arranged as to produce a triangular quantum well in the narrow gap material. Figure 7 shows the conduction band for such a structure where the wide gap material is heavily *n*-doped. This "modulation doped" arrangement causes free carriers from the heavily doped material to fall into the confining potential of the nominally undoped material. This structure provides the exciting possibility of introducing carriers into a narrow channel with very high mobility. This high mobility arises from the fact that the ionized width provides important scattering centers at low temperatures which are isolated from the free carriers. This isolation can be increased further by introducing an undoped spacer layer on the wide bandgap side. The triangular well behaves in a similar way to the square quantum well. It causes quantization of the energy levels in the direction perpendicular to the junction, in a way that resembles the Si-SiO_2 interface of an Si MOSFET [1]. The wave functions are not the standing wave solutions in Equation (3.1); the two lowest states are described by the approximate forms

$$\varphi_0(x) = 2a^{2/3}xe^{-ax} \tag{4.1}$$

$$\varphi_1(x) = Nx(1 - Bx)e^{-bx} \tag{4.2}$$

where a and b are variational parameters; N is anormalization constant and B is chosen such that φ_0 and φ_1 are orthogonal. In the directions parallel to the junction the carriers are still unrestricted. Usually only the lowest subband is occupied at T = 0 K; i.e., there is a "freeze out" of one degree of freedom. In this sense the system is really 2D. The restricted freedom of the carriers results in fewer states being available for scattering, leading to a reduced scattering rate and consequently, higher mobility. Mobilities approaching the room temperature mobility of intrinsic GaAs have been reported and mobilities in excess of $10^6 cm^2/Vs$ have been measured at low temperatures [18]. The major application of the heterojunction are based around the high electron mobility transistors (HEMT, MODFET, or TEGFET) and related devices.

Figure 8 schematically shows a single GaAs/GaAlAs heterojunction field effect transistor (FET). The advantage of this structure is that the concentration of the 2D electron gas can be varied continuously by means of

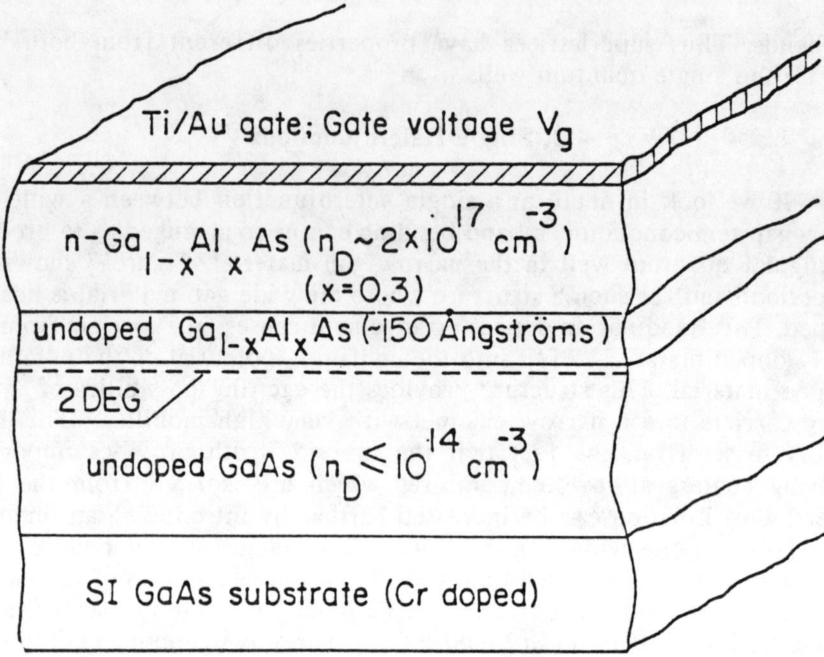

Fig. 8. A schematic diagram of the GaAs/GaAlAs heterojunction FET used to create a high-mobility 2D electron gas.

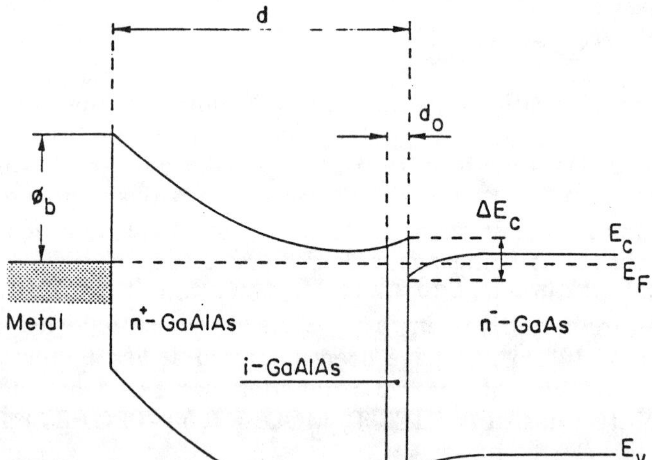

Fig. 9. Band diagram through the GaAs/GaAlAs heterojunction FET (Figure 8).

an applied gate voltage, typically from $\approx 10^{11}$ cm^{-2} to zero. This is illustrated in Figure 9.

The structure in Figure 8 can be modified so that 1D subbands are also obtained. Figure 10 shows how this is achieved by means of a split gate [19, 20[. By applying a negative gate voltage V_g the electrons beneath the gated regions may be removed, leaving a conducting channel under the slit. The widths of this channel may be varied, typically from 1 μm to zero by varying V_g. For a narrow channel, effects of 1D quantization become important. Suppose the lowest perpendicular state $\varphi_0(x)$ in Equation (4.1) is occupied. For a wide channel the system is then effectively 2D. With decreasing width of the conducting channel the separation of the 1D sublevels increases. Therefore, only the lowest 1D subband is occupied in the limit of a very narrow channel; i.e., there is a freeze-out of two degrees of freedom. In this sense the system is truly 1D. Thus, the split gate GaAs/GaAlAs heterojunction FET offers an elegant method for studying 2D → 1D dimensional crossover. We will return to the experimental observation of these 1D subbands in a later section. Other ways of inducing 1D subbands in ultranarrow Si inversion layers and GaAs/GaAlAs mesa structures are described in Refs. [21, 22].

4.23. A Special GaAs Field Effect Transistor (GaAsFET)

The split gate heterojunction FET discussed above allows us to study the 2D → 2D dimensional crossover. The special GaAsFET, originally proposed by Pepper [23] (and Poole et al. [24]), allows similar studies in the case of 3D → 2D crossover. As shown in Figure 11, the device consists of a homogeneously n-doped active layer, grown on a p^+ substrate via a nominally undoped GaAs region. An ohmic contact to the substrate enables a substrate depletion region to be set up, screening the conducting electrons in the n-region from any interface states and enabling conduction in a region of constant doping. A metal contact defines the gate depletion region. The gate bias V_g. In practice, the channel width ranges between a few thousand Angstroms and pinch-off. Both the 3D and 2D limits are therefore accessible in the same device.

The junction field effect transistor (JFET) discussed by Maan [25] serves the same purpose. In a GaAs pnp structure consisting of alternately grown thin layers of n and p type doping in the same host material, the electrons from the donors transfer to the acceptors in the p layers. The resulting space charge leads to a potential well, confining the electrons in the center of the n layer in the same way as in the GaAsFET above. Through selective contacts to the n and p layers, the potential difference

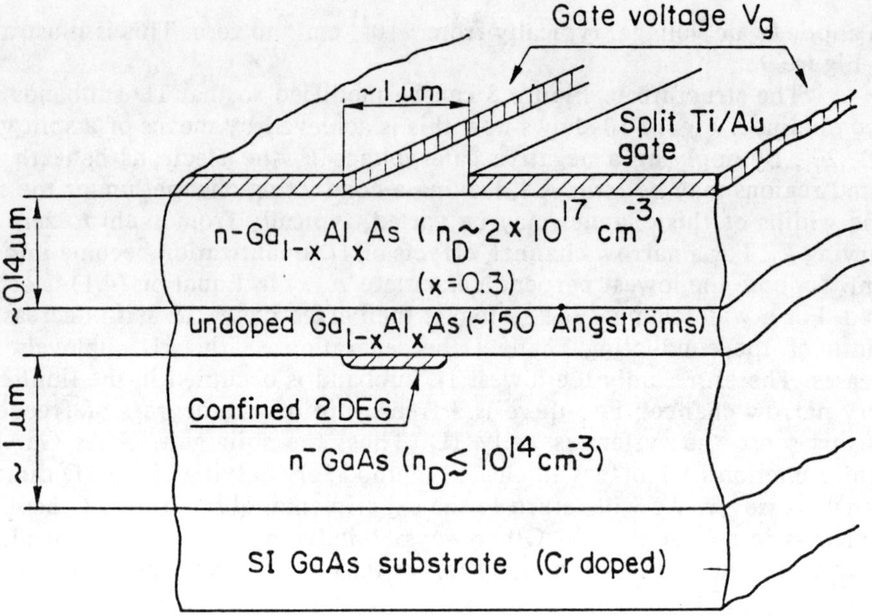

Fig. 10. A schematic illustration of the split gate GaAs/GaAlAs heterojunction FET used to define a narrow channel in a 2D electron gas.

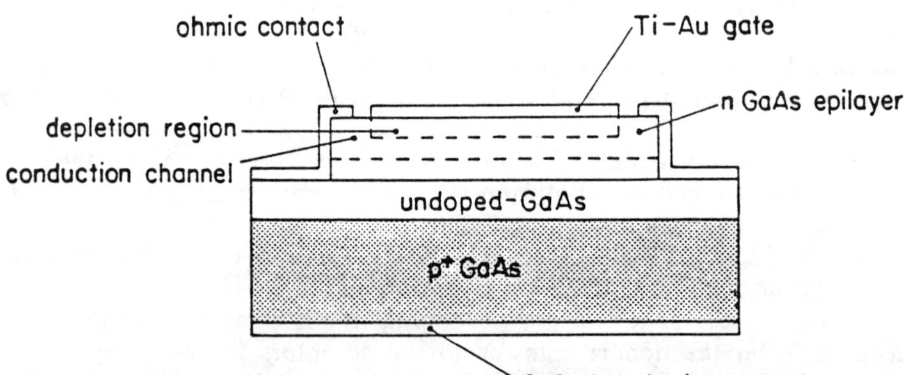

Fig. 11. Cross-section through the special GaAs FET.

between these layers can be varied externally. In this way the real carrier concentration in the n layer can be modulated over a wide range.

The shape of the confining well, shown schematically in Figure 12, can be modeled quite easily if we first make the simplifying assumption that the random arrangement of ionized point impurities may be smeared into a uniform positive background. If n_D is the number of donors/unit volume and $\rho(x)$ is the electron density, the potential energy $V(x)$ is then determined by the one-dimensional Poisson equation

$$V''(x) = \frac{e^2}{\kappa}[n_D - \rho(x)] \tag{4.3}$$

(The static screening κ is 12.1 ε_0 in the case of GaAs). The general solution of Equation (4.3) is, apart from a constant defining the reference energy,

$$V(x) = \frac{n_D e^2}{2\kappa} x^2 - \frac{e^2}{\kappa}\int_0^x (x - x')\rho(x)dx' \tag{4.4}$$

where the center of the symmetric channel defines the origin of the coordinate system. If we now adopt the simple approximation that the electrons are also uniformly smeared over a neutral channel, as in the abrupt depletion model [27], we find a potential energy varying as

$$V(x) = \frac{m^* \omega_0^2}{2}(|x| - t/2)^2 \tag{4.5}$$

for $|x| \geqslant t/2$. For $|x| < t/2$ we have $V(x) = 0$. The "characteristic" frequency is $\omega_0 = e\sqrt{n_D / m^* \kappa}$. The expression for the thickness of the channel follows from the boundary conditions,

$$t = \sqrt{\frac{2\kappa}{en_D}(\Phi_b - V_p)} - \sqrt{\frac{2\kappa}{en_D}(\Phi_b - V_g)} \tag{4.6}$$

where Φ_b is the Schottky barrier as shown in the band diagram of Figure 12 and V_p is the pinch-off voltage when the gate and substrate depletion regions meet. From the neutrality condition we find the real concentration of electrons,

$$N_e = n_D t \tag{4.7}$$

The simple modeling above that has frequently been employed, turns out to be quite accurate. A more exact quantum mechanical treatment [26] gives

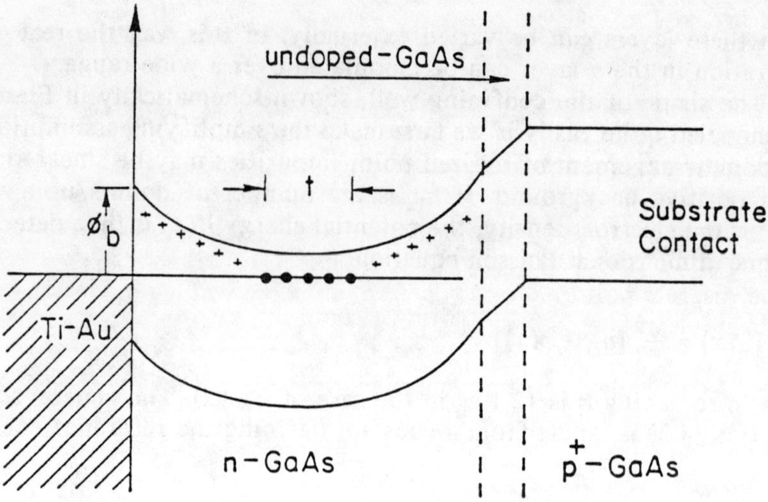

Fig. 12. Electrostatic potential through the GaAs FET. Electrons are confined to a conducting channel of width t.

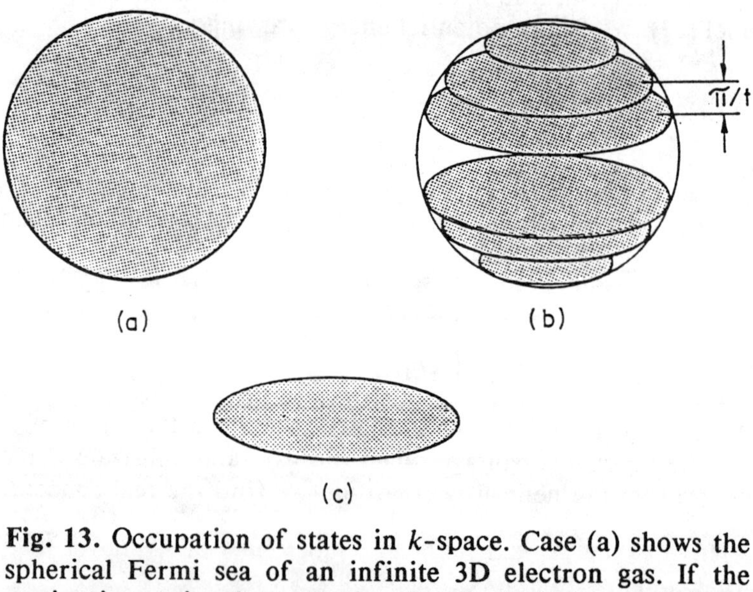

Fig. 13. Occupation of states in k-space. Case (a) shows the spherical Fermi sea of an infinite 3D electron gas. If the motion is restricted to one direction (by infinite steep walls) 2D subbands are formed. The Fermi sphere is sliced into discs defining the occupancy in k-space (b). Case (c) refers to the case of only the ground 2D subband being occupied.

results which are numerically very close to our simple abrupt depletion model.

For a very wide channel, let us say a few thousand Angstroms, the electron gas is in practice 3D. The potential is dominated by its flat interior region and the exact shape of the confining walls matters less. This situation also persists when we make the channel somewhat narrower. We should therefore expect that the quantized sublevels are approximately the same as for an infinitely deep square well, Equation (3.2), and that the wave vector takes the discrete values $k_x = n\pi/t$. As the channel is made finite, we may therefore think of the original 3D Fermi sphere as being sliced as shown in Figure 13. As t decreases the spacing between the 2D "Fermi discs", π/t, increases. As a result discs are successively expelled from the interior of the 3D sphere, that is, electrons are removed from the channel as dictated by the neutrality condition. At the same time the Fermi energy, E_F, relative to the ground sublevel, drops in a monotonous way below its 3D free electron value. At a sufficiently small value of t only one Fermi disc remains occupied. We refer to this gradual depopulation of the subbands as 3D → 2D dimensional crossover. WE may think about the 2D → 1D crossover previously mentioned in a similar way.

The infinitely deep square well may serve as an illustration of the depopulation of subbands with decreasing t. Quantitatively it has, however, less to offer in the limit of narrow channels. For example, when $t \rightarrow 0$, Equation (4.5) shows that the confining potential is a parabola. The sublevels are then the same as for a quantum harmonic oscillator,

$$E_n = \hbar\omega_0(n + 1/2) \tag{4.8}$$

with $n = 0, 1, \ldots$ Fortunately, the limits of very narrow and wide channels can be bridged by means of the semiclassical WKB method [24]. Using this method for the potential in Equation (4.5) we find the analytical result

$$E_n = \left\{ -\frac{\omega_0 t}{2\pi}(2m^*)^{1/2} + \left[\hbar\omega_0(n + 1/2) + 2m^* \left(\frac{\omega_0 t}{2\pi} \right)^2 \right]^{1/2} \right\}^2 \tag{4.9}$$

For $t \rightarrow 0$ we recover Equation (4.8) from this equation. Also in the limit $t \rightarrow \infty$ the correct expression is found.

By means of the expression for E_n in Equation (4.9) the Fermi energy E_F relative to the ground sublevel E_0 may now be found. Requiring that E_F must be common to all occupied subbands we have

$$(E_n - E_0) + \hbar^2 k_{F\parallel,n}^2/2m^* = E_F \tag{4.10}$$

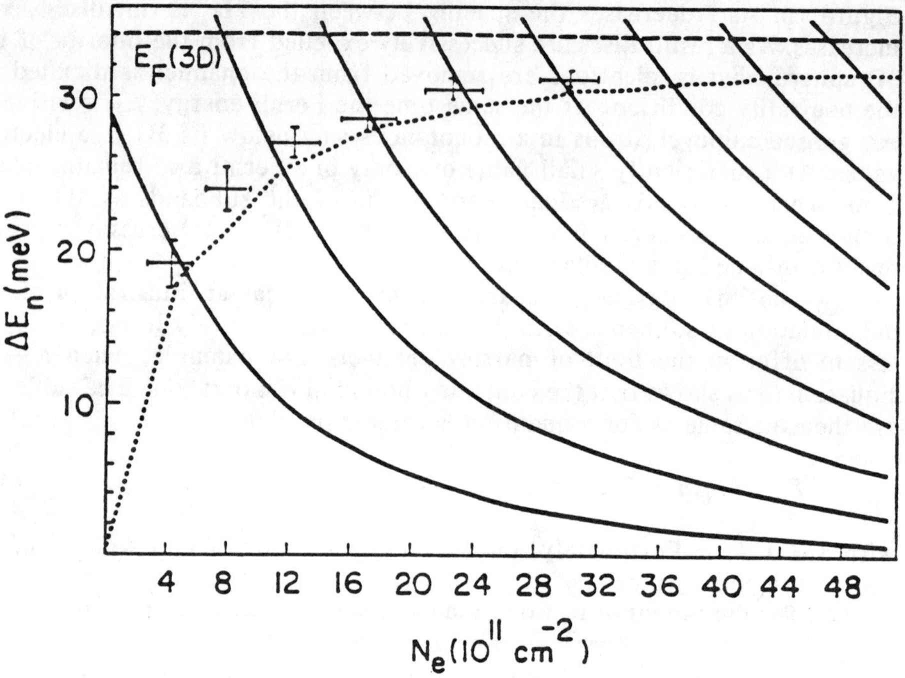

Fig. 14. Sublevel quantization and Fermi energy relative to the ground sublevel as a function of a real electron concentration in a GaAsFET with n_D = $5{\cdot}10^{17}$ cm^{-3}. Solid curves refer to the sublevel separations ΔE_n according to a semiclassical expression in Equation (4.9), and the dotted curve to the corresponding Fermi energy E_F. The straight horizontal dashed line shows the position of the Fermi energy E_F(3D) for a uniform noninteracting 3D electron gas. Experimental values of E_F obtained from magneto-resistance oscillations, are given with error bars (from Ref. [24]).

The Fermi wave number $k_{F\parallel,n} = [2\pi N_{n,e}]^{1/2}$ is associated with the motion along the channel; $N_{n,e}$ is the density of electrons occupying level n ($\Sigma N_{e,n} = N_e$). Figure 14 shows the sublevels in a GaAsFET and how E_F gradually drops below its 3D value as the channel width decreases. The same figure also shows experimental estimates of E_F derived from magneto-resistance oscillations in a magnetic field parallel to the channel [24]. Our theoretical characterization of the GaAsFET suggests, together with experimental results, that the idea of a gradual dimensional crossover because of QSE is indeed working in practice.

In conclusion, we have described how QSE can be realized in a number of small semiconductor structures. In our review we have been quite selective. Thus, we have not discussed the important case of Si MOSFETs by which the physics of 2D electron gas was pioneered [1], δ-doping layers in GaAs [28, 29], Si/Si$_x$Ge$_{1-x}$ superlattices [30, 31], etc. The outline is, however, general enough to serve as a first guidance to QSE in these systems as well.

5. *Effects of QSE on Electrical Transport*

Electrical conduction is an issue of scattering. The discussion above refers to an idealized situation in which effects of scattering on subbands and density of states are ignored. This may be warranted for high-mobility structures like modulation doped GaAs/GaAlAs heterojunctions, in which the ionized impurity scatterers are spatially separated from the carriers. The case of the GaAsFET discussed above is, on the other hand, different. Here the conducting channel resides right in the middle of randomly distributed ionized impurities, dielectrically screened by the carriers. Scattering is therefore expected to be stronger in this case. Our assumption of a uniformly smeared background of ionized donors is therefore bound to be a rough one. At low temperatures there are three regions of behavior in the device:

1) 3D metallic conduction;

2) 2D quasi-metallic conduction;

3) 2D thermally activated conduction among strongly localized states.

The random scattering potential, the difference between the real potential and the model potential in Equation (4.5), is the origin of electrical resistance. For a very wide channel the dielectric screening from the large number of carriers is relatively effective in reducing the strength of the scatterers. The system therefore behaves as a 3D metallic system, somewhat like a simple liquid metal. In a more narrow channel the screening is weaker, but still strong enough to support metallic conduction. Due to the increasing importance of certain multiple scattering events resulting in weak localization (see Refs. [4, 32]) corrections to the usual Drude behavior appear. There is an unusual low-temperature dependence of the conductivity; the system is quasi-metallic. For a very narrow channel, finally, the screening is insufficient because of the small number of carriers. The electrons therefore localize because of a strong, random potential.

For a quasi-metallic system it is reasonable, as a first approximation, to regard the scattering potential as a weak perturbation allowing low order perturbation theory to be applied. The standard approach to electrical transport is to use the Boltzmann equation. The Boltzmann conductivity, which results from such treatment, is also the starting point for including effects of weak localization. The localized regime requires, of course, a different approach.

An analysis of the weak scattering case leads to a number of interesting points regarding subbands. In a situation in which several occupied subbands are occupied, the standard Boltzmann transport equation must be modified [33, 34]. Let the electrical field E be parallel to the channel (along the z-axis). At steady-state the collision term has to balance the term induced by the external field. For a carrier with wave vector $k_{\parallel}$ in the nth subband we thus have [34]:

$$\sum_{n',k} w(nk_{\parallel}; n'\vec{k}'_{\parallel})(f(n',k'_{\parallel}) = eE\frac{\delta}{\delta p_z}f_0(n,k_{\parallel}) \tag{5.1}$$

where p_z is the z-component of the parallel momentum $\vec{p} = \hbar k_{\parallel}$. Here $w(nk;$ $\vec{n}'_{\parallel}k')$ is the transition rate for scattering from a state $|nk_{\parallel}\rangle$ to $|n'k'_{\parallel}\rangle$ and *vice versa*, $f(n,k_{\parallel})$ is the distribution at steady-state, and $f_0(n,k_{\parallel})$ is the initial zero-field Fermi distribution function. Before the field is turned on the occupied states are defined by a set of concentric discs centered at $k_{\parallel} = 0$ as in Figure 13 (the arguments can readily be extended to quasi 1D-systems). Equation (5.1) is solved with the assumption that the Fermi disc associated with each subband is rigidly displaced by an amount Δk_z^n because of the electrical field, i.e., $f_n(n,k_{\parallel})$ is set equal to the displaced initial Fermi distribution

$$f_0(n, k_y, k_z + \Delta k_z^n) \simeq f_0(n,k_{\parallel}) + \hbar\Delta k^n{}_z\delta f_0(n,k_{\parallel})/\delta p_z \tag{5.2}$$

in Equation (5.1) and similarly for $f(n',k'_{\parallel})$. The displacements Δk_z^n for the different subbands need not be identical. In fact, the assumption of a displacement common to all the occupied subbands eventually leads to non-physical results. Introducing a transport time τ_n for each occupied subband by the relation $\hbar\Delta k^n{}_z = eE\tau_n$ we find at zero temperature, using Equation (5.1) and assuming elastic scattering:

$$A_n\tau_n + \sum_{n'} B_{nn'}\tau_{n'} = 1 \tag{5.3}$$

where

$$A_n = \sum_{n'} \sum_{k'} w(nk;\, n'k'_{\parallel})$$

$$B_{nn'} = - \sum_{k'} w(nk;\, n'k'_{\parallel})k'_z/k_z \tag{5.4}$$

Here the summation over n' extends over all occupied subbands and all wave vectors refer to the Fermi level. The number of Equations (5.3) equals the number of occupied subbands. Hence there is in general a corresponding number of transport times τ_n. In the Boltzmann approximation the sheet conductance is consequently,

$$g_B = \frac{e^2}{m*} \sum_{n}^{occ} N_{n,e}\tau_n \tag{5.5}$$

In the present model transport times and conductance are not smooth functions of electron concentration, but are strongly affected by the presence of subbands. It then appears that QSE would make it possible to directly observe subbands by measuring conductance. This argument is made more transparent if we for a moment assume that the static scatterers are s-wave scatterers, for example, δ-function and other short range scatterers. Such a simplification will also reveal the shortcomings of the present Boltzmann model as we will see shortly. For s-wave scatterers the coefficients B_{nn}' in Equation (5.4) are identically equal to zero. Then $\tau_n^{-1} = A_n$, that is, transport times and finite lifetimes due to disorder scattering are equal. To evaluate τ_n we now have to specify the transition rate. In general w is given by the "Golden Rule" expression

$$w(nk_{\parallel};\, n'k'_{\parallel}) = (2\pi/\hbar)|\langle nk_{\parallel}|V_s|n'k'_{\parallel}\rangle|^2\delta(E_nk_{\parallel}) - E_n'k'_{\parallel})) \tag{5.6}$$

where V_s is the scattering potential. $|\langle nk_{\parallel}|V_s|n'k'_{\parallel}\rangle|$ is a configurationally averaged quantity which is independent$k_{\parallel}$ and $k'_{\parallel}$ for s-scatterers. Thus, from Equation (5.4)

$$\tau^{n1} = (2\pi/\hbar) \sum_{n'} |\overline{V_{nn'}}|^2 \int \frac{dk}{(2\pi)^2} \delta(E_F - E_{n'}(k_{\parallel})) \tag{5.7}$$

where $V_{nn'}$ are the simplified $k_{\parallel}$-independent matrix elements which need no further elaboration. The integral in the expression above is, however, just the definition of the density of states for the nth subband, $g^n_{2\text{D}}(E)$ in Equation (3.6). Therefore,

$$\tau_2^{-1} \propto \sum_{n'} |V_{nn'}|^2 g_{2D}^{n}(E_F) \tag{5.8}$$

But $g_{2D}^{n}(E_F)$ is a step function, as discussed in Section 3. Thus each time E_F passes a sublevel E_n there will be a discontinuous change in τ_n. This means that the conductance will display a strong structure as the width of the channel is varied. The physics behind this is that each time E_F passes into an empty subband on increasing t, new scattering channels are abruptly opened. Hence the discontinuous increase in resistance. Figure 15 shows typical theoretical results for the sheet resistance of a GaAsFET [33] assuming dielectrically screened point impurities, i.e., a more realistic choice than the δ-function scatterers. Low-temperature experimental results are also shown in the same figure. Obviously the sharp structure in resistance as predicted by the Boltzmann model is not confirmed by experiments [33, 35], although weak traces of subbands have been seen in the differential resistance [36, 37]. There may be different reasons for this failure. One reason would be a poor quality of the device and therefore a badly defined conducting channel. As we will see below, however, there is an inherent weakness in the theoretical treatment itself.

In the arguments above we have assumed that the scattering takes place between states unperturbed by disorder; $k_{\parallel}$, for example, is a good quantum number. The density of states is highly idealized in the same way. This may be an accurate description if the mobility is exceedingly high. For most devices, in particular our GaAsFET, the situation is, however, different. Because of the scattering due to disorder $k_{\parallel}$, for example, is not a good quantum number but there will be a spread over different $k_{\parallel}$-vectors. Rather than a single wave there will be a wave packet, the width of which is characterized by the lifetime. This blurring is easily incorporated in the time-dependent perturbation theory leading to the "Golden Rule" expression in Equation (5.7). The usual procedure is to replace the sharp δ-function in Equation (5.7) by a Lorentzian

$$\delta(x) \rightarrow \frac{1}{\pi} \frac{1}{x^2 + \Gamma^2} \tag{5.9}$$

where $\Gamma = \hbar/2\tau$. With lifetime broadening incorporated we therefore have, instead of Equation (5.9), the self-consistent set of equations

$$\Gamma_n = \sum_{n'} \overline{|V_{nn'}|^2} \frac{dk'_{\parallel}}{(2\pi)^2} \frac{\Gamma_{n'}}{(E - E_n(k'_{\parallel}))^2 + \Gamma_{n'}^2} \tag{5.10}$$

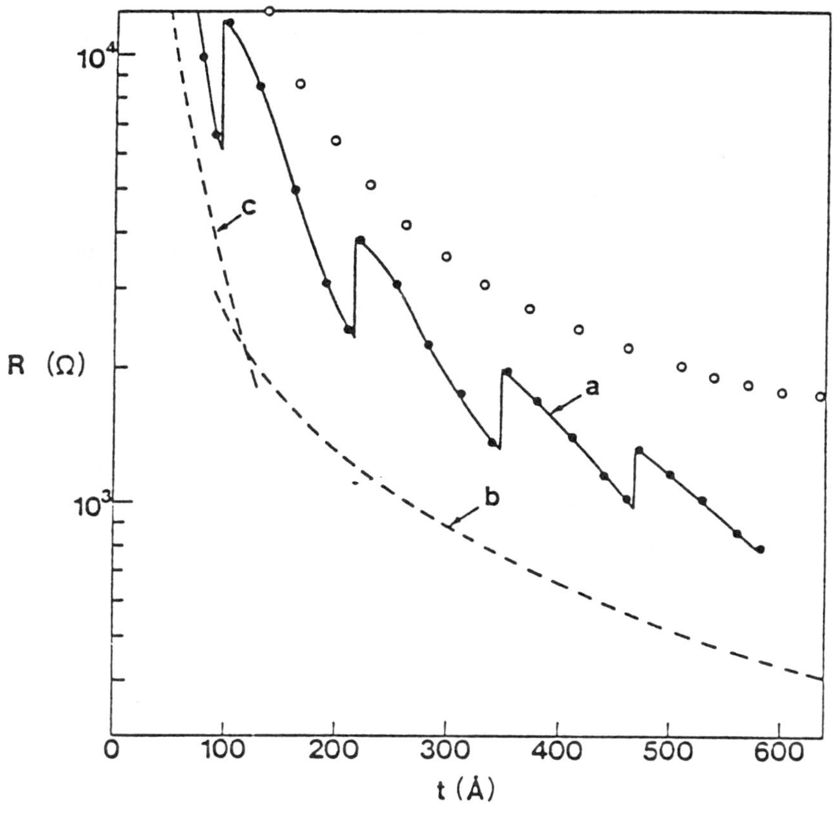

Fig. 15. Sheet resistance R versus channel thickness in a GaAsFET with $n_D = 5{\cdot}10^{17}$ cm^{-3}. Full circles represent actual results from the Boltzmann transport equation applied to random Coulomb scatterers dielectrically screened by the carriers occupying the conduction band. Subbands are included at all stages of the calculations. The full curve (a) is a free interpolation between calculated points. Curves (b) and (c) are 3D and 2D asymptotes, respectively. The open circles refer to the measurements at 4.6 K (from Ref. [33]).

(As $\Gamma_n \to 0$ this equation reduces to Equation (5.7)). Once more the integral in this expressions is simply the 2D density of states modified by a finite lifetime due to disorder, $\tilde{g}^n_{2D}(E)$. In contrast to the sharp features of our previous density of states $\tilde{g}^n_{2D}(E)$ is now a smooth function. The discontinuous steps in the total density of states are therefore blurred. Depending on the strength of the scattering potential, and the level separation ΔE we encounter different situations. If $\Delta E > \hbar/\tau$ a smooth structure is still retained, but for $\Delta E < \hbar/\tau$ all structure in the density of states is removed (Figure 16).

When lifetime broadening is included in the description of electron states and density of states right from the start, the conductance cannot be derived in the same elementary way as in the case of the Boltzmann transport theory above. Instead we must resort to a Kubo formula using Green's function methods. Thus, expressions for the conductance in quasi 2D and 1D systems at low temperatures have been derived in this way [38, 39]. The derivation is somewhat technical and we shall not give any details here. Qualitatively the effects of lifetime broadening are as would be intuitively expected (Figure 17); QSE in the conductance gradually disappears with decreasing mobility and/or sublevel separations.

The picture of lifetime broadening gives a good explanation of the negative results for our GaAsFET. If we now take a look at some other semiconductor structures with higher mobilities than the GaAsFET, the general trend is consistent with the Cantrell-Butcher theory [38]. Thus, subband related structure has been observed in an $In_{0.53}Ga_{0.47}As$ JET [37] at low temperatures of 4.2 K. The reason for the enhanced mobility is a lower effective mass. At higher temperatures, 77 K, QSE disappears because of the thermal occupation of higher subbands. Thermal blurring thus acts in a way that is siilar to finite lifetimes. QSE is also clearly observed in the case of a modulation doped, high-mobility GaAs/GaAlAs heterojunction. The mobility shows a pronounced decrease at the transition from one-subband to two-subband conduction [40]. These examples single out the role of mobility. Still must require that $\Delta E > \hbar/\tau$. If we consider, for example, the split gate GaAs/GaAlAs heterojunction FET in Figure 10, we are obviously dealing with a high-mobility system. In spite of this a subband related structure is not observed, since a depopulation of the 1D subbands is induced by varying the gate voltage, the reason being that the 1D sublevel separations are too small (~1-2 MeV). This point brings us to another method for enhancing the observability of QSE. By applying a suitably oriented magnetic field the level separations can be diamagnetically increased to the extent that the condition $\Delta E > \hbar/\tau$ is met. The argument brings Subnikov-de Haas oscillations to mind. To observe SdH we require $\omega_c\tau \gg 1$, ω_c being the cyclotron frequency.

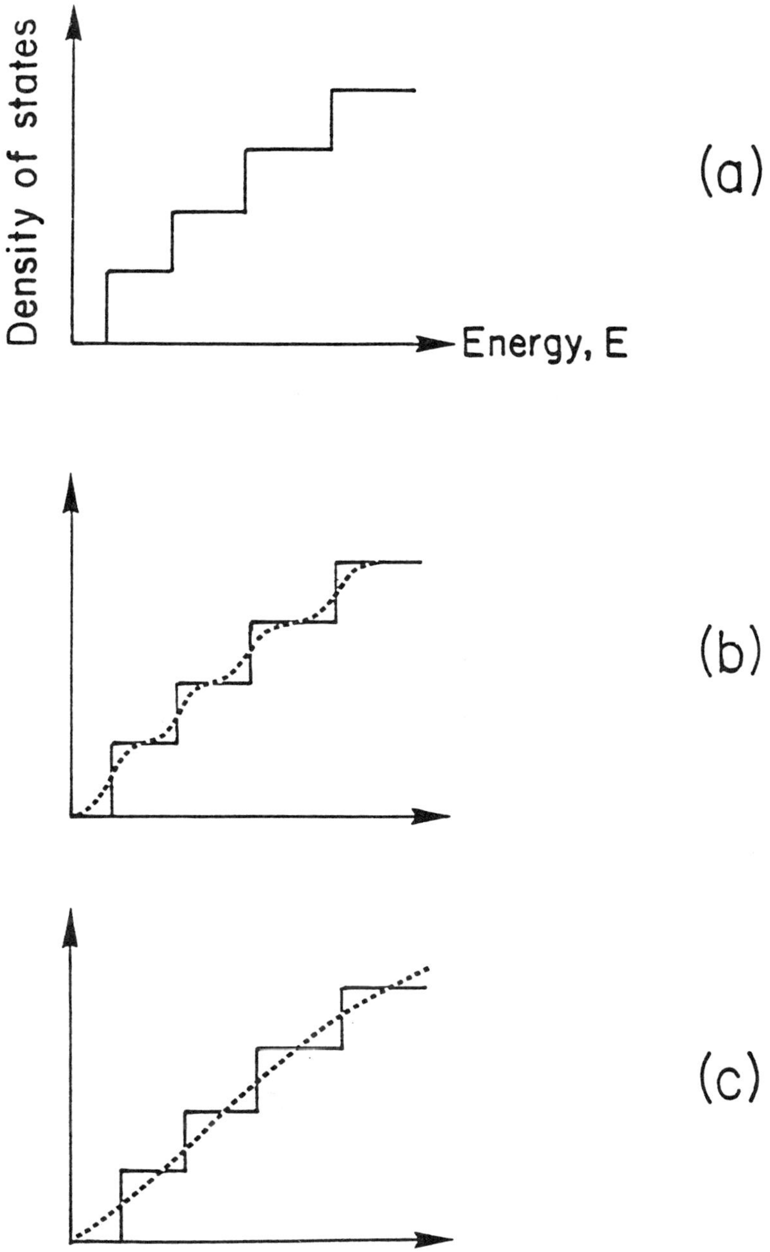

Fig. 16. Density of states broadened by a finite lifetime: (a) initial, unperturbed case, (b) $\Delta E > \hbar/\tau$, (c) $\Delta E \approx \hbar/\tau$.

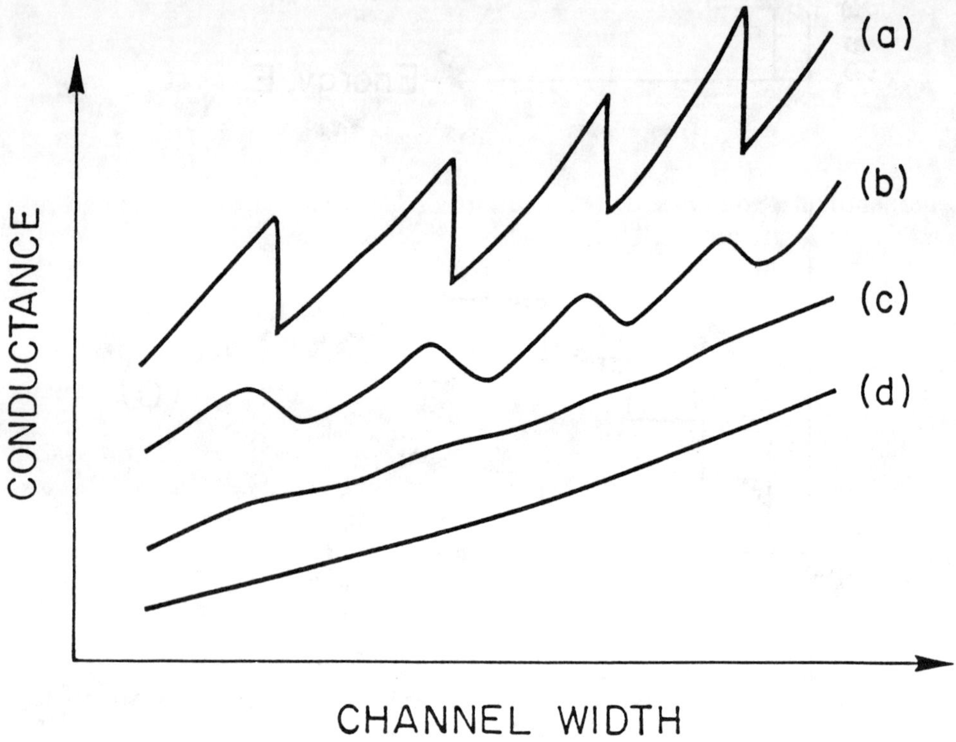

Fig. 17. Schematic representation of conductance at $T = 0$ K and zero magnetic field in a FET or JFET with several occupied subbands. Curve (a) refers to the Boltzmann transport equation modified to a multisubband situation. Lifetime broadening is not incorporated in this model. Each time E_F passes into a new subband on increasing the channel width there is a discontinuous drop in the conductance. Curves (b)-(d) show the effects of lifetime broadening as scattering due to disorder grows stronger (decreasing mobility).

6. Magnetic Depopulation of Subbands

Let us consider the GaAsFET discussed above. In the case of a magnetic field $\vec{B}$ applied parallel to the conducting channel, a magnetic term must be included in the Schrödinger equation in addition to the electrostatic confinement, equation (4.5). The magnetic field is defined as $\vec{B}$ = curl $\vec{A}$ and choosing the Landau gauge $\vec{A} = (0, Bx, 0)$ gives the field in the z-direction (see Ref. [41]). The magnetic term to be added to the Hamiltonian is then

$$V_B(x) = \frac{m^* \omega_c{}^2}{2} (x - x_0)^2 \tag{6.1}$$

where $x_0 = p_y/(eB)$ and $\omega_c = eB/m^*$: $p_y = \hbar k_y$ is the momentum component perpendicular to the field. The solutions of the Schrödinger equation are still in the form given by Equation (3.3). If there is no electric confinement, $\xi_n(x)$ is a Landau state centered at $x = x_0$. Equation (6.1) also applies to the split gate heterojunction FET. However, the magnetic field should now be chosen to be perpendicular to the surface of the 2D electron gas.

When both electrical and magnetic confinement are present, a variety of different cases arise depending on the position of $V_B(x)$ (see Figure 18). In a strong magnetic field and for small values of $|k_y|$ the magnetic parabola is well inside the constant region of the electrostatic potential. To a good approximation the corresponding solutions are then Landau states. For larger values of $|k_y|$, however, an electron also starts to interact more strongly with one of the electrostatic confining walls. Because of this additional squeezing, the energies of these "edge states" are gradually raised with increasing $|k_y|$. Consequently, the initial parabolic dispersions in Equation (3.4) are severely distorted by the applied magnetic field as shown in Figure 19. At the same time the density of states acquires sharper features. We will not describe the full solution of this problem here but refer to the detailed treatment in Ref. [42]. Instead, we will consider a simplified model which is easily solved exactly. Although this model strictly applies only to a very narrow FET channel, it illustrates the mechanism of magnetic depopulation of subbands in a transparent way.

Let the electrostatic confinement be defined by the parabolic from

$$V(x) = \frac{m^* \omega_0^2}{2} x^2$$

The sublevels are then as in Equation (4.8).

Adding the two parabolas $V(x)$ and $V_B(x)$ we find for the total confinement the shifted parabola

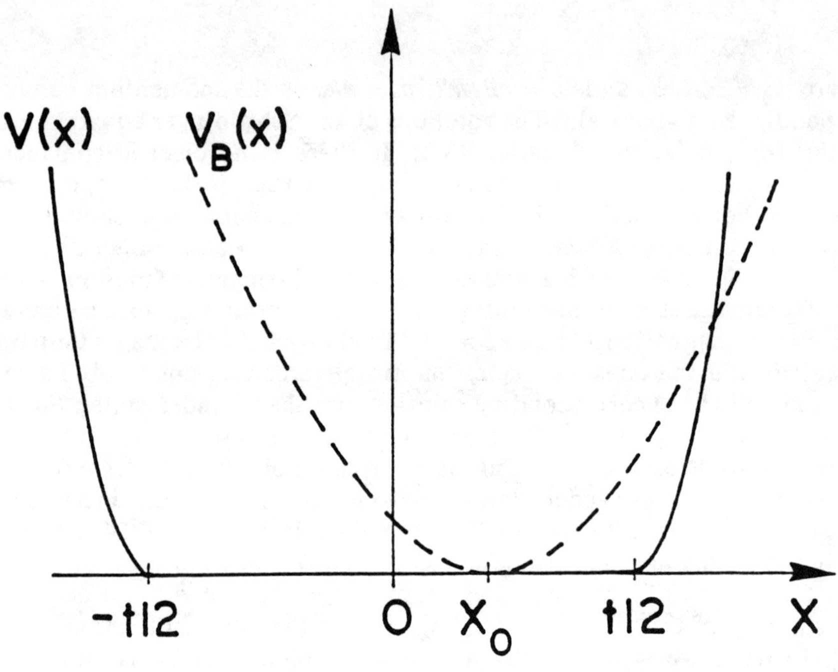

Fig. 18. Schematic view of the confining potentials. $V(x)$ is the electrostatic channel potential in Equation (4.5) (full curve) and $V_B(x)$ the magnetic parabola, Equation (6.1), centered at $x = x_0$ (broken curve).

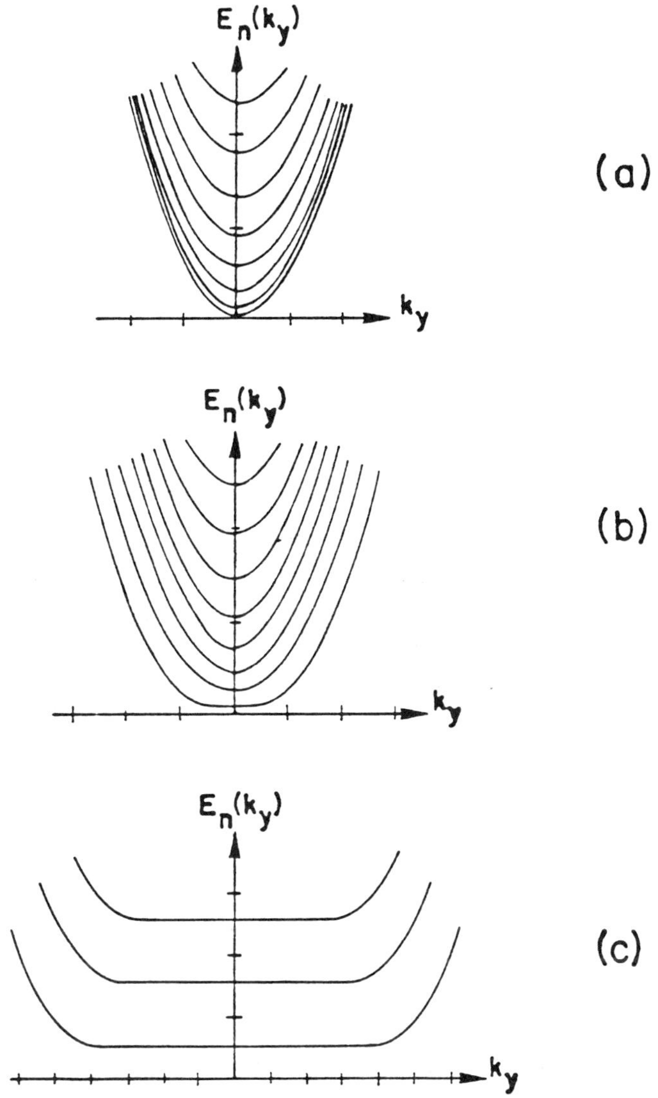

Fig. 19. Subbands for electrons confined by the potential in previous figure. The dispersions $E_n(k_y)$ refer to motion perpendicular to the magnetic field ($k_z = 0$). (a) shows the case of zero magnetic field, (b) refers to a moderate field, and (c) a strong magnetic field. Dispersionless regions correspond to Landau states centered well inside the confining walls, while the edge states show strong dispersion.

$$V(x) + V_B(x) = \frac{m^*\omega^2}{2}(x - \bar{x}_0)^2 + \frac{\hbar^2 k_y^2}{2m^*(B)} \tag{6.3}$$

where $\bar{x}_0 = x_0(\omega_c/\omega)^2$, $\omega = \sqrt{\omega_0^2 + \omega_c^2} > \omega_0$ and $m^*(B) = m^*(\omega/\omega_0)^2 > m^*$ is an effective "magnetic mass." The eigenenergies corresponding to this harmonic oscillator model are

$$E_n(k_y) = E_n(B) + \frac{\hbar^2 k_y^2}{2m^*(B)} \tag{6.4}$$

Here $E_n(B) = \hbar\omega(n + \frac{1}{2})$ which would define 1D sublevels in, for example, the split gate heterojunction FET. The magnetic field thus acts in two ways: the separation of the sublevels, $\Delta E = \hbar\omega$, increases and the electrons appear heavier, resulting in a flattening of the subband dispersions. Because ΔE increases with B the subbands successively depopulate as illustrated in Figure 20. This process is accelerated by the simultaneous change in the density of states, Equations (3.11) and (3.12), in which m^* and E_{nl} are to be replaced by $m^*(B)$ and $E_n(B)$. The density of states thus becomes more sharply peaked in the presence of a magnetic field with a greater possibility of observing structure in the conductance due to subband depopulation.

For the 2D subbands in a GaAsFET (or JEFET) the translational motion along the z-axis, the direction of the magnetic field, must be added to equation (6.4). The one-electron dispersions are therefore

$$E_n(k_\parallel) = E_n(B) + \frac{\hbar^2 k_y^2}{2m^*(B)} + \frac{\hbar^2 k_z^2}{2m^*}, \tag{6.5}$$

i.e., we still have parabolic dispersions but the effective mass appears anisotropic. The corresponding density of states is obtained from Equation (3.6) by inserting $E_n = \hbar\omega\ (n + \frac{1}{2})$ for E_n and multiplying the entire expression by ω/ω_0. We therefore find the same mechanism as in the quasi 1D case: depopulation takes place with increasing magnetic field because ΔE increases and the density of states is enhanced. Therefore, oscillations in the resistance should now appear, provided that the magnetic field is strong enough. Although these oscillations remind us of Shubnikov-de Haas oscillations, they are distinctly different. For example, a $1/B$ plot of the resistance maxima against level index n does not result in a straight line as does SdH maxima. Actually, departure from a straight line behavior can be taken as a marker of geometrical confinement.

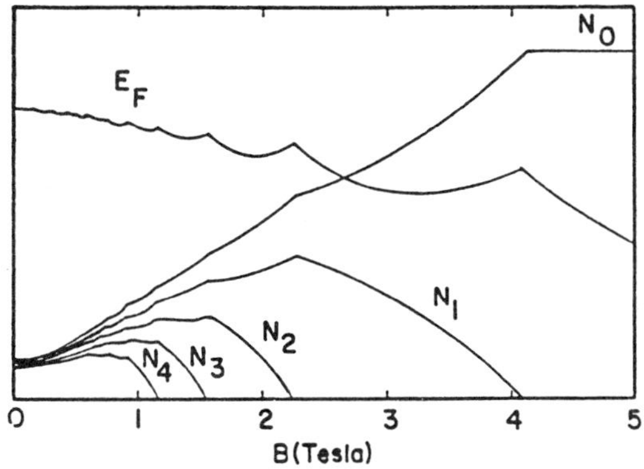

Fig. 20. Magnetic depopulation of 1D subbands in a harmonic confining potential, Equation (6.2). N_i is the number of electrons/m occupying the ith subband, and E_F is the Fermi energy relative to the ground sublevel. In the calculations ω_0 and the total electron concentration were set equal to 10^{12} s^{-1} and $7{\cdot}10^{12}$ m^{-1}, respectively. For these values $E_F \simeq 9$ MeV at $B = 0$. (G. Roos and K.-F. Berggren, unpubl.).

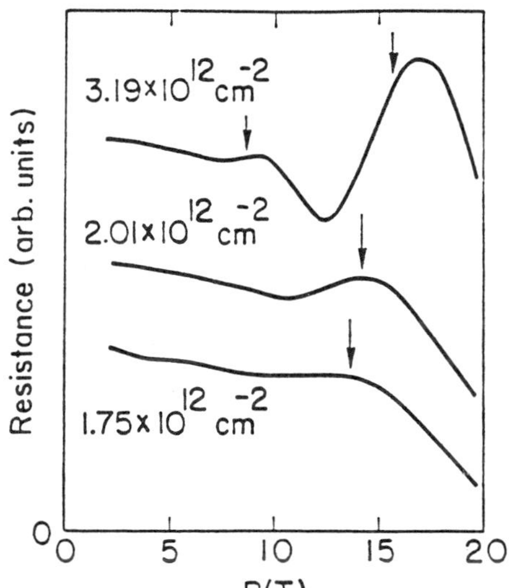

Fig. 21. Resistance vs. magnetic field in a GaAsJFET with $n_D = 7{\cdot}10^{17}$ cm^{-3} at different real electron concentrations/channel widths (Ref. [25]). The magnetic field is parallel to the conducting channel. Arrows indicate theoretical estimates of the magnetic field at which a magnetic depopulation of the subbands takes place (Ref. [42]).

Figure 21 shows the measured resistance in a GaAsJFET with $n_D = 7{\cdot}10^{17}$ cm^{-3} (Refs. [24, 25]). In contrast to the previous case of the GaAsFET in zero magnetic field a structure is now clearly observed. Semiclassical WKB calculations [42] of the magnetic depopulation using the more accurate confining potential in Equation (4.5) combined with the magnetic parabola in Equation (6.1), result in magnetic fields which correlate well with the measured structure as shown in Figure 21.

In the previous section we have mentioned the split gate GaAs/GaAlAs heterojunction FET and the failure of observing 1D subbands as the width of the channel was varied by a gate voltage. As for the GaAsJET this situation is drastically changed when a perpendicular magnetic field is applied. Strong oscillations which correlate with 1D subbands are now observed [20].

In summary we have demonstrated how QSE becomes observable by means of a suitably oriented magnetic field. The primary reason is that the diamagnetic separation of the sublevels makes it possible to satisfy the condition $\Delta E > \hbar/\tau$ by choosing the field strong enough. Secondly, the depopulation is accelerated by the enhancement of the density of states. We have explicitly discussed a GaAsJFET and a split gate GaAs/GaAlAs heterojunction. The method of magnetic depopulation is, however, generally applicable as shown by measurements on, for example, $In_{0.53}Ga_{0.47}As$/ $In_{0.52}Al_{0.48}As$ heterojunctions [43] (depopulation of two or three 2D subbands), $In_{0.53}Ga_{0.47}$/InP heterojunctions [44] (two 2D electrical subbands), and δ-doping layers in GaAs [28] (multi-2D subbands). A number of interesting theoretical problems need attention. At present there is no accurate theory for the megnetotransport in geometrically confined systems like the zero-field theory of Cantrell and Butcher [38]. Consequently, there are uncertainties about exactly which feature of the measured resistance is to be associated with subband depopulation. Furthermore, effects of weak localization and interactions and the associated quantum fluctuations in multi-subband systems require more theoretical studies. In many cases, for example, in various quasi 1D devices, a more realistic modeling of the (electrostatic) confining potential should also be performed.

7. QSE in Some Other Low-Dimensional Systems

We have described QSE occurring in small semiconductor structures and devices at some length. QSE is, of course, not unique to these systems although the small effective mass and carrier concentration make it easier to satisfy the quantization conditions in practice. Thus, there have been considerable experimental and theoretical studies of QSE in thin films of metals and semimetals, in particular Bi. The early literature is reviewed in Ref.

[45]. The field is still active as indicated by Refs. [46, 47]. Qse is also expected in another context, namely, the onset of superconductivity and superfluidity. Thus the superconducting transition temperature in Sn films has been found to oscillate with film thickness. The oscillations have been attributed to QSE but some problems about the periodicity remain unclear. QSE has also been studied theoretically in connection with superfluid ^{3}He films [49]. In the thickness regime $k_F^{-1} \ll t < \xi$, where ξ is the coherence length, the transition temperature is depressed as the thickness t is decreased and exhibits a weak oscillatory behavior. Similar features are also to be expected for other systems undergoing phase transitions because of pairing.

8. Conclusion

This introduction to quantum aspects of small semiconductor structures and devices can only serve as an entry point to part of a vast and rapidly growing literature. The field has attracted attention not only because of the usefulness of these systems in applications - both present and proposed - but also because they provide relatively simple realizations of many physical phenomena previously confined to textbooks, as well as unexpected new effects. The intense activity around the world in devising new structures and looking for new phenomena promises more to come.

Acknowledgements

I am indebted to the organizers of the "Kandy Symposium on Condensed Matter Physics" for the opportunity to participate in this series of lectures. I also thank the International Centre for Theoretical Physics, Trieste, Italy, and the Knut and Alice Wallenberg Foundation for financial support.

References

1. Ando, T., Fowler, A. B., Stern, F., *Rev. Mod. Phys.*, vol. 54, p. 437, 1982.
2. Bauer, G., Kuchar, F., Heinrich, H. (eds.), *Two-Dimensional Systems, heterostructures, and Superlattices*, Springer, 1984.
3. Bauer, G., Kuchar, F., Heinrich, H. (eds.), *Two-Dimensional Systems: Physics and New Devices*, Springer, 1986.
4. Kelly, J. J., Weisbuch, C. (eds.), *The Physics and Fabrication of Microstructures and Microdevices*, Springer, 1986.

5. *Physics Today*, Oct 1986.
6. Prange, R. E., Girvin, S. M. (eds.), *The Quantum Hall Effect*, Springer, 1987.
7. Merlin, R., Bajema, K., Clarke, R., Juang, F.-Y., Bhattacharya, P. K., *Phys. Rev. Lett.*, vol. 55, p. 1768, 1985.
8. Datta, S., Bandyopadhyay, S., *Phys. Rev. Lett.*, vol. 58, p. 717, 1987, and references cited.
9. Kohn, W., In: *Solid State Physics*, ed. by. F. Seitz and D. Turnbull, vol. 5, Academic Press, 1957.
10. Abram, R. A., Rees, G. J., Wilson, B. L. H., *Adv. Phys.*, vol. 27, p. 799, 1978. (A review article with an extensive list of references).
11. Berggren K.-F., Sernelius, B. E., *Phys. Rev. B*, vol. 24, p. 1971, 1981; *Solid State Electronics*, vol. 28, p. 11, 1985.
12. Sakaki, H., *Japan J. Appl. Phys.*, vol. 19, p. L735, 1980; *J. Vac. Sci. Technol.*, vol. 19, p. 148, 1981.
13. Kubakaddi, S. S., Mulimani, B. G., *J. Appl. Phys.*, vol. 59, p. 3643, 1985; *J. Phys. C: Solid State Phys.*, vol. 18, p. 6647, 1985.
14. Kubakaddi, S. S., Mulimani, B. G., Jali, V. M., *J. Phys. C: Solid State Phys.*, vol. 19, p. 5453, 1986.
15. Watanabe, H., in Ref. 4, p. 5.
16. Watanabe, H., Inoshita, T., *Optoelectron. Dev. Tech.*, vol. 1, p. 33, 1986.
17. Inoshita, T., Ohnishi, S., Oshiyama, A., *Phys. Rev. Lett.*, vol. 57, p. 2560, 1986.
18. Weimann, G., Schlapp, W., in Ref. 3, p. 33, and references cited.
19. Thornton, T. J., Pepper, M., Ahmed, H., Andrews, D., Davies, G. J., *Phys. Rev. Lett.*, vol. 56, p. 1198, 1986.
20. Berggren, K.-F., Thornton, T. J., Newson, D. J., Pepper, M., *Phys. Rev. Lett.*, vol. 57, p. 1769, 1986.
21. Warren, A. C., Antoniadis, D. A., Smith, H. I., *Phys. Rev. Lett.*, vol. 17, p. 1858, 1986.
22. van Houten, H., van Wees, B. J., Heijman, M. G. J., André, J. P., *Appl. Phys. Lett.*, vol. 49, p. 1781, 1986.
23. Pepper, J., *J. Phys. C: Solid State Phys.*, vol. 10, p. L173, 1977.
24. Poole, D. A., Pepper, M., Berggren, K.-F., Hill, G., Myron, H. W., *J. Phys. C: Solid State Phys.*, vol. 15, p. L121, 1982.
25. Maan, J. C., in Ref. 2, p. 183.
26. Berggren, K.-F., Newson, D. J., *Semicond. Sci. Tech.*, vol. 1, p. 246, 1986.
27. See, for example, S. M. Sze, *Physics of Semiconductors*, 2nd ed., Wiley, 1982.
28. Zrenner, A., Reisinger, H., Koch, F., Ploog, K., Maan, J. C., *Phys. Rev. B*, vol. 33, p. 5607, 1986.

29. Koch, F., Zrenner, A., Zahana, M., in Ref. 3, p. 175.
30. Abstreiter, G., Brugger, H., Wolf, T., Jorke, H., Herzog, H. J., *Phys. Rev. Lett.*, vol. 54, p. 22, 1985.
31. Abstreiter, G.,m Brugger, H., Wolf, T., Zachai, R., Zeller, C., in Ref. 3, p. 130.
32. See, for example, the extensive review by P. A. Lee and T. V. Ramakrishnan, *Rev. Mod. Phys.*, vol. 57, 287, 1985.
33. Sernelius, B. E., Berggren, K.-F., Tomak, M., McFadden, C. M., *J. Phys. C: Solid State Phys.*, vol. 18, p. 225, 1985.
34. Roth, L. M., Argyres, P. N., *Semiconductors and Semimetals*, Vol. 1, eds. Willardson and Beer, Academic Press, 1966, p. 159.
35. Newson, D. J., McFadden, C. M., Pepper, M., *Phil. Magnetic. B*, vol. 52, p. 437, 1985.
36. Berggren, K.-F., Widlund, B. L., *Int. J. Quant. Chem.*, vol. 19, p. 567, 1986.
37. Newson, D. J., Ph.D. Thesis, Cambridge Univ., UK, 1986.
38. Cantrell, D. G., Butcher, P. N., *J. Phys. C: Solid State Phys.*, vol. 18, pp. 5111, 6639, 1985.
39. Kearney, M. J., Butcher, P. N., *J. Phys. C: Solid State Phys.*, vol. 20, p. 47, 1987.
40. Störmer, H. L., Gossard, A. C., Wiegmann, W., *Solid State Commun.*, vol. 41, p. 107, 1982.
41. Landau, L. D., Lifshitz, E. M., *Quantum Mechanics*, 2nd ed., Pergamon, 1985.
42. Berggren, K.-F., Newson, D. J., *Semicond. Sci. Tech.*, vol. 1, p. 327, 1986.
43. Newson, D. J., Berggren, K.-F., Pepper, M., Myron, H. W., Davies, G. J., Scott, E. G., *J. Phys. C: Solid State Phys.*, vol. 19, p. L403, 1986.
44. Gouldner, Y., Vieren, J. P., Voos, M., Delahaye, F., Dominguez, D., Hirtz, J. P., Razeghi, M., *Phys. Rev. B*, vol. 33, p. 3990, 1986.
45. Tavger, D. A., Demikhovskii, *Usp. Fiz. Nauk.*, vol. 96, p. 61, 1969.
46. Batra, I. P., Ciraci, S., Srivastava, G. P., Nelson, J. S., Fonga, C. Y., *Phys. Rev. B*, vol. 34, p. 8246, 1986.
47. Dryzek, J., Czapla, A., *Phys. Rev. Lett.*, vol. 58, p. 721, 1987.
48. Orr, B. C., Jaeger, H. M., Goldman, A. M., *Phys. Rev. Lett.*, vol. 53, p. 2046, 1984.
49. Tesanovic, Z., Valls, O. T., *Phys. Rev. B.*, vol. 33, p. 3139, 1986.

TOPICS IN THE THEORY OF ELECTRON TRANSPORT FOR QUASI-2D AND QUASI-1D ELECTRON GASES

P. N. Butcher
University of Warwick
Coventry, CV4 7AL, UK

Abstract: **When the effect of the scatterers is neglected the density of states of a quasi-2D (1D) electron gas exhibits step discontinuities (square root singularities). This produces sharp quantum size effects in both the electrical conductivity and diffusion thermopower calculated from Boltzmann's transport equation. When the effect of the scatterers on the density of states is taken into account the step discontinuities (square root singularities) are smoothed off, as is the structure in the transport coefficients, it remains significant even when the mean free path is in the order of the reciprocal of the Fermi wavenumber f particularly for a quasi-1D gas. The phonon drag contribution to thermopower is evaluated for a 2D electron gas interacting with 3D acoustic phonos at low temperatures.**

1. Introduction

Experiments on the electrical conductivity (σ) of systems with narrow conducting channels — both metal films and semiconductor systems (inversion layers, heterostructures, quantum wells, etc.) have been carried out by many researchers (for example, Ogrin et al. 1966; Poole et al. 1981). In both classes of experiment, discontinuities in the conductivity have been observed as the channel width or Fermi energy is changed. These discontinuities have been called quantum-size effects (QSE) as they arise from discontinuities in the electronic density of states. When the width of the channel is comparable to the de Broglie wavelength of the electrons, electron motion perpendicular to the channel is no longer free. A subband structure develops and it is this structure that gives rise to the QSE. However, the discontinuities observed experimentally have not been as sharp as predicted by theory and in some cases QSE have not been observed at all (see, for example, Sernelius et al. 1985).

Recent thermopower (S) experiments on GaAs/AlGaAs heterostructures have produced very different zero magnetic field results (see Fletcher et al. 1986; Davidson et al. 1986). The results of Fletcher et al. (1986) show S to have both a magnitude (of the order of 300 μVK^{-1} at 5 K) and a temperature variation which suggest the dominant contribution is from the phonon drag mechanism in agreement with theory (Cantrell and Butcher 1986). This conclusion is in broad agreement with measurements on Si MOSFETs see, for example, Zavaritskii and Kvon 1983; Gallagher et al. 1986). However, the results of Davidson et al. appear to be rather puzzling; here S has a magnitude (5 μVK^{-1}) and temperature variation below ~10 K consistent with the conclusion that S is due purely to the diffusion mechanism. Above 10 K S departs strongly from the linear temperature variation predicted by the usual analysis of the diffusion mechanism (see, for example, Blatt 1968) and decreases sharply with a temperature exhibiting a minimum at approximately 20-30 K. This behavior is unexpected and draws some exotic explanations from authors. Cantrell (1986) has shown that the data are consistent with the diffusion mechanism over almost the entire temperature range if it is postulated that the first excited subband lies several MeV above the Fermi energy.

The description of electron transport in these narrow-channel systems has been largely based on the Boltzmann equation (see, for example, Siggia and Kwok, 1970). This predicts pronounced QSE as the Fermi energy enters a new subband. The smoothing of these discontinuities in practice has been attributed to the broadening of the electron states due to scattering (SErnelius et al. 1985, Mori and Ando 1979). In reviewing the theoretical interpretation of these measurements we begin in Section 2 with a discussion of the effect of level broadening on QSE in σ. The Section starts with a brief discussion of the calculation of σ from Boltzmann's transport equation and ends with an account of a recent extension of the theory for a quasi-2D electron gas (i.e., one confined quantally to a thin film) to the case of a quasi-1D electron gas (i.e., one confined quantally to a fine wire). given by Kearney and Butcher (1986, 1987). In Section 3 we discuss diffusion thermopower and in Section 4 we outline the theory of 3D phonon drag on a 2D electron gas recently developed by Cantrell and Butcher (1987).

2. The Influence of Level Broadening on QSE in the Electrical Conductivity of Quasi-2D and Quasi-1D Electron Gases

2.1. Introduction

In this paper we calculate the electrical conductivity of a quasi-2D and a quasi-1D electron gas when it is limited by impurity scattering. The approximations inherent in the Boltzmann equations are exposed and we are able to modify the derivation of the conductivity to allow for broadening of the electronic states.

First we consider a system of electrons with effective mass m^*, confined in the z direction but free in the xy plane, interacting with a randomly positioned assembly of scattering centers. Electron-electron interactions will not be included. The hamiltonian of this system is given by

$$H = -\frac{\hbar^2}{2m^*}\left(\frac{\partial^2}{\partial x^2} + \frac{\partial^2}{\partial y^2}\right) + \sum_i u(r - r_i, z - z_i) - \frac{\hbar^2}{2m^*}\frac{\partial^2}{\partial z^2} + V(z) \quad (1)$$

where $r = (x, y)$ and we have written the z coordinate explicitly to emphasize the geometry of the system. $V(z)$ is the potential confining the electrons in the z direction. When the scattering potential $u = 0$ we see that the energy eigenfunctions are

$$\psi_{ak} = A^{-1/2}\phi_\alpha(z)e^{ik\cdot r} \quad (2a)$$

which has the energy

$$\varepsilon_\alpha(k) = \varepsilon_\alpha + \hbar^2(k_x^2 + k_y^2)/2m^*. \quad (2b)$$

Here, the $\phi_\alpha(z)$ are solutions of the 1D Schrödinger equation

$$-(\hbar^2/2m^*)(\partial^2/\partial z^2)\phi_\alpha(z) + V(z)\phi_\alpha(z) = \varepsilon_\alpha\phi_\alpha(z) \quad (3)$$

where ε_α is the associated energy. In the simplest case $V(z)$ has infinite walls at $z = 0$ and t and is zero in between (a one-dimensional "box") so that

$$\phi_\alpha(z) = (2/t)^{1/2}\sin(\alpha\pi z/t) \tag{4a}$$

and

$$\varepsilon_\alpha = (\hbar^2/2m^*)(\pi/t)^2\alpha^2 \tag{4b}$$

In (2a) A denotes the area of the system in the xy-plane.

We see that the dispersion relation (2b) takes the form of a series of "subbands" labeled by the index α which we give the values 1, 2, ... The energy in subband α is given by (2b), ε_α is the subband minimum and the energy in the subband increases from there as $\hbar^2/2m^*$ times the square of the 2D k-vector (k_x, k_y). When only subband 1 is occupied we have a strictly 2D electron gas. When several subbands are occupied we have a *quasi*-2D electron gas.

2.2. The Conductivity at 0 K Predicted by Boltzmann's Equation

We assume that the electron scattering in subband α may be described by a momentum relaxation time $\tau_\alpha(\varepsilon)$ which is a function of the energy ε alone. Then the usual treatment of conductivity at 0 K for a 3D band (Blatt 1968; Butcher 1973) applies for each subband. Thus we obtain

$$\sigma(\varepsilon_F) = \sum_\alpha n_\alpha e^2\tau_\alpha/m^* \tag{5}$$

where n_α is the 2D electron density in subband α when the Fermi level is ε_F.

It is easily seen that the density of states is equal to $m^*/\pi\hbar^2$ each subband for $\varepsilon > \varepsilon_\alpha$ and vanishes for $\varepsilon < \varepsilon_\alpha$. Consequently, n_α in (3) is proportional to $(\varepsilon_\alpha - \varepsilon_F)\theta(\varepsilon_\alpha - \varepsilon_F)$, where $\theta(\varepsilon_F - \varepsilon_\alpha)$ is the unit step function which varies continuously with ε_F. On the other hand τ_α is in the order of the inverse lifetime of the state αk. This may be calculated by using Fermi's Golden Rule and is proportional to the density of final states available for transitions from αk. This increases discontinuously by $m^*/\pi\hbar^2$ whenever ε_F passes through a subband minimum. Consequently, τ_α discharges discontinuously and so does $\sigma(\varepsilon_F)$. We show the variation of σ with ε_F in Figure 1.

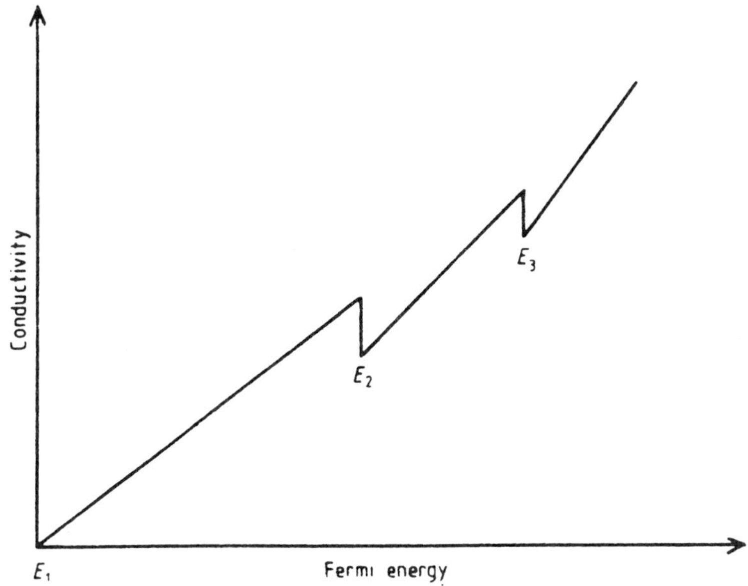

Fig. 1. Schematic plot of the zero-temperature conductivity against Fermi energy. E_1, E_2, and E_3 are the energy minima of the first three subbands.

2.3. The Effect of Level Broadening

The analysis given in Section 2.2 assumes that the energy levels are sharp. In fact they are broadened by the scattering potential in (1). The primary effect of level broadening is to round off the "staircase" structure found above for the density of state.

To take this effect into account approximately, we use the "general Born approximation" (Cantrell and Butcher 1985a). The density of states is written in the form

$$N(\varepsilon) = \frac{2}{A\pi} \sum_{\alpha} \sum_{k} \operatorname{Im} G_{\alpha\alpha}(k) \tag{6}$$

where $G_{\alpha\alpha}(k)$ is the $(\alpha k, \alpha k)^{\text{th}}$ matrix element of the Green's operator in the unperturbed energy representation given in Section 2.1. We may write

$$G_{\alpha\alpha}(k) \simeq \left[\varepsilon - \varepsilon_{\alpha}(k) + i\,\Gamma_{\alpha\alpha}(\varepsilon)\right]^{-1} \tag{7}$$

where

$$\Gamma_{\alpha\alpha}(t) = \frac{1}{2\tau_\alpha(\varepsilon)} \tag{8}$$

with $\tau_\alpha(\varepsilon)$ denoting the lifetime of a state with energy ε in subband α. (For the δ-function scattering potential considered below there is no distinction between $\tau_\alpha(\varepsilon)$ in (8) and (5). For this reason we use the same notation for both quantities.)

When (7) is substituted into (6) we may evaluate the sum over k by taking the limit $A \to \infty$ for which

$$\frac{1}{A}\sum_k \to 4\pi^2 \int dk \tag{9}$$

the integral over the 2D k space is readily evaluated to yield

$$N(\varepsilon) = \sum_\alpha \frac{m^*}{\pi\hbar^2}\cdot\frac{1}{\pi}\left[\frac{\pi}{2} + \tan^{-1}\frac{\varepsilon - \varepsilon_\alpha}{\Gamma_{\alpha\alpha}(\varepsilon)}\right] . \tag{10}$$

We see by inspection of (10) that, when $\Gamma_{\alpha\alpha}(\varepsilon) \to 0$, the contribution to $N(\varepsilon)$ from subband α reduces to 0 when $\varepsilon < \varepsilon_\alpha$)so that the inverse tangent $\to -\pi/2$) and reduces to $m^*/\pi\hbar^2$ when $\varepsilon > \varepsilon_\alpha$ (so that the inverse tangent $\to +\pi/2$). Thus we regain the staircase density of states function assumed in Section 2.2 for the zeroth order case. To go beyond this it is necessary to make a self-consistent calculation of $\Gamma_{\alpha\alpha}(\varepsilon)$. This is easy to do in only one case. We suppose that the scattering potential $u(r - r_i, z - z_i)$ in (11) is a 3D δ-function with strength U, i.e.,

$$u(r - r_i, z - z_i) = U\,\delta(x - x_i)\delta(y - y_i)\delta(z - z_i). \tag{11}$$

Then we may use (8) and Fermi's Golden Rule transition rates to write an expression for $\Gamma_{\alpha\alpha}(t)$ involving transitions to each subband with a strength proportional to the contribution of that subband to the total density of states. A self-consistent equation for the $\Gamma_{\alpha\alpha}(\varepsilon)$ is obtained when the latter is determined from (10) with a factor 2 removed because the scattering potential does not flip spins. Thus we obtain (Cantrell and Butcher 1985a):

$$\Gamma_{\alpha\alpha}(t) = n_s \sum_\beta \overline{M_{\alpha\beta}}^2 \frac{m^*}{\pi\hbar^2}\cdot\frac{1}{\pi}\left[\frac{\varepsilon - \varepsilon_\beta}{\Gamma_{\beta\beta}(\varepsilon)}\right] \tag{12}$$

where n_s is the 2D density of scatterers and

$$M_{\alpha\beta}{}^2 = U^2 \int dz_i P(z_i)\phi^2(z_i)\phi^2(z_i) \tag{13}$$

is the mean square matrix element associated with an individual scatterer located at (x_i, y_i, z_i). In (13) $P(z_i)$ denotes the probability density for z_i. In the following we take $P(z_i)$ to be t^{-1} over the channel width t and 0 outside the channel. In writing (12), the probability distribution of r_i has been assumed to be uniformly distributed over A.

Equation (12) has been solved numerically after using the subband wavefunctions (4a) to evaluate $M_{\alpha\beta}{}^2$ (Cantrell and Butcher 1985a).

The above calculation of the effect of level broadening on the density of states may be used to develop a heuristic formula for $\sigma(\varepsilon_F)$ in the presence of level broadening. To do so we start from the elementary formula (5) and rewrite it in the suggestive form

$$\sigma(\varepsilon_F) = \sum_\alpha \frac{m^*}{\pi\hbar^2} \theta(\varepsilon_F - \varepsilon_\alpha)(D_\alpha(\varepsilon_F) \tag{14}$$

where

$$D_\alpha(\varepsilon_F) = (\varepsilon_F - \varepsilon_\alpha)\tau_\alpha(\varepsilon_F)/m^* \tag{15}$$

has the immediate interpretation as the electron diffusivity in subband α. The coefficient of $D_\alpha(\varepsilon_F)$ in (14) is the zeroth order approximation to the density of states in subband α. To improve the formula we simply replace this by the better approximation derived by inspection of Equation (10). Thus we have

$$\sigma(\varepsilon_F) = \sum_\alpha \frac{m^*}{\pi\hbar^2} \left[\frac{1}{2}\left(\frac{\pi}{2} + \tan^{-1}\frac{\varepsilon_F - \varepsilon_\alpha}{\Gamma_{\alpha\alpha}(\varepsilon_F)}\right) \cdot D_\alpha(\varepsilon_F)\right]. \tag{16}$$

There is a problem with (16). Equation (15) implies that $D_\alpha(\varepsilon_F)$ is negative when $\varepsilon_F < \varepsilon_\alpha$. Hence subband energies in this range make a negative contribution to σ. This result is physically wrong. To correct it requires a full diagrammatic calculation of $\sigma(\varepsilon_F)$ (Cantrell and Butcher 1985a). We do not give the details here because the final result turns out to be a very simple modification of (16). It is only necessary to addressed 1 to the expression in square brackets. Thus we obtain the final formula for $\sigma(\sigma_F)$ in the general Born approximation:

$$\sigma(\varepsilon_F) = \sum_\alpha \frac{e^2}{2\pi^2} \left[\left(\frac{\pi}{2} + \tan^{-1}\frac{\varepsilon_F - \varepsilon_\alpha}{\Gamma_{\alpha\alpha}(\varepsilon_F)}\right)\frac{\varepsilon_F = \varepsilon_\alpha}{\Gamma_{\alpha\alpha}(\varepsilon_F)} + 1\right] \tag{17}$$

where $\tau_{\alpha\alpha}(\varepsilon_F)$ is to be determined by solving (12).

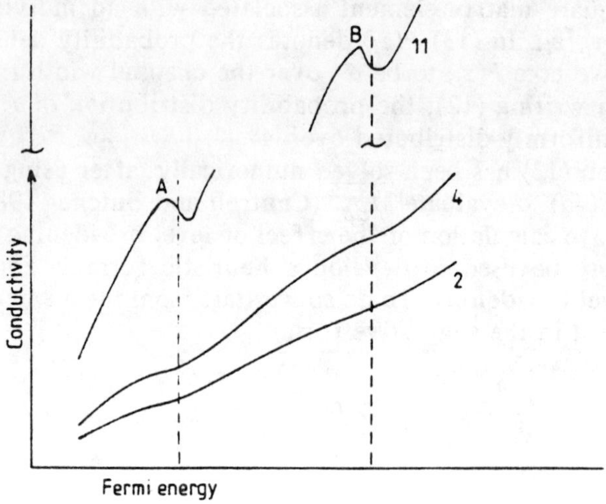

Fig. 2. The variation of conductivity Fermi energy for several approximate values of $(k_F l)_{2D}$ given by the curves. (The depth of the dip at A is $\simeq$6% of the highest value before the dip and the dip at B has a depth $\simeq$3% of the highest previous value.) The dashed lines indicate the positions of the subband minima.

The final results of the calculation are shown in Figure 2. For each curve we start with the Fermi level in the lowest subband so that the system is strictly 2D. The number labeling the curves is the value of $k_F l$ (i.e., $k_F l_{2D}$) in that case where l is the mean free path and k_F is the radius of the Fermi circle. Each curve is generated by increasing k_F with $k_F l$ held fixed. We see that strong fine structure remains when $k_F l \sim 11$ but that it has almost disappeared when $k_F l \sim 2$. The latter value is close to that obtained in the experiments Sernelius et al. (1985).

2.4. The Effect of Level Broadening on the Density of States Conductivity of a Quasi-1D System

The calculations carried out above for a quasi-2D electron gas are readily extended to a quasi-1D electron gas (Kearney and Butcher 1987). In (1) $V(z)$ is replaced by a 2D well $V(y, z)$. Consequently, the electrons are confined in both the y and z directions. They remain free only in the x direction and we are concerned with a "quantum wire."

In (2) $k = (k_x, 0)$ and α becomes a double index which labels the localized states associated with $V(y, z)$.

The QSE are stronger for a quasi-1D gas than they are for a quasi-2D gas because the step discontinuities in the zeroth order density or states for the latter case are replaced by square root singularities. The effect of level broadening on the density of states is illustrated in Figure 3 where $(k_F l)_{1D}$ has an interpretation similar to $(k_F l)_{2D}$ in Figure 2.

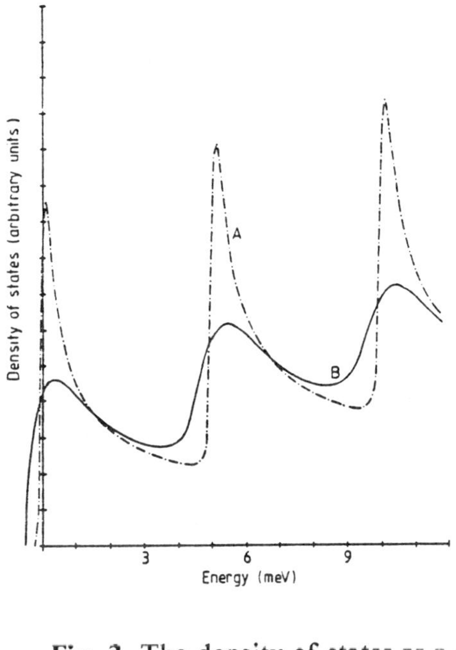

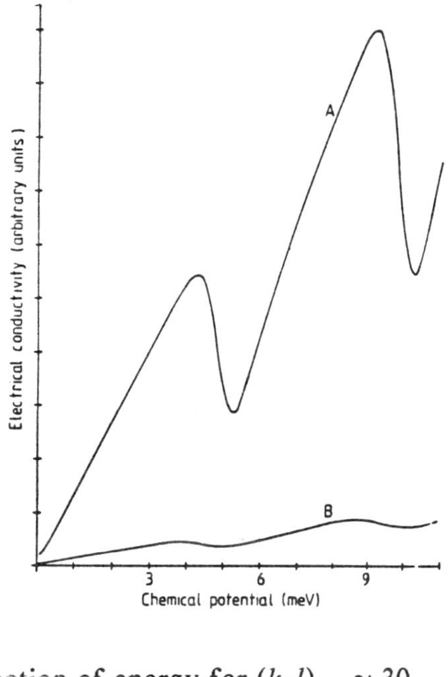

Fig. 3. The density of states as a function of energy for $(k_F l)_{1D} \simeq 30$ (A), 5 (b). The subband minima are at 0.5 and 10 MeV.

Fig. 4. The electrical conductivity as a function of chemical potential for $(k_F l)_{1D} \simeq 30$ (A), 5 (B), at a temperature of $T = 2$ K. The subband minima are at 0.5 and 10 MeV.

The corresponding effect on $\sigma(\varepsilon_F)$ is shown in Figure 4 for the same values of $k_F l$. Systems of this type can be fabricated using MBE techniques but QSE in them have yet to be investigated experimentally (Maines 1987).

3. The Influence of Level Broadening on QSE in the Diffusion Thermopower of Quasi-2D and Quasi-1D Electron Gases

3.1. Introduction

We consider an isotropic medium of simplicity. Then the linear macroscopic transport equations may be written in the form

$$E' = \rho J + S\nabla T \tag{18a}$$

$$Q = \pi J - \kappa\nabla T \tag{18b}$$

Here E' and ∇T denote the emf and temperature gradient respectively while J and Q are the electric current density and heat flux. The scalar coefficients in (18) are the electrical resistivity $\rho = \sigma^{-1}$, where σ is the electrical conductivity, the thermopower S, the Peltier coefficient π and the thermal conductivity κ (Butcher 1973). These are the quantities which are measured in transport experiments. Resistivity is the most commonly measured transport coefficient. Electronic thermal conductivity has been of little interest in recent years but many studies have been made of the thermopower. We wee from (18a) that, to obtain S, it is only necessary to measure the emf and temperature gradient along a bar-shaped sample which carries no electric current and take their ratio. This is the way *measurements* of S are usually carried out. However, an alternative route may be followed because of the time-reversal symmetry of the system under consideration. This gives rise to the Onsager symmetry relation $\pi = TS$ (Butcher 1973). We see from (18b) that S is therefore T^{-1} times the ratio of heat flux to electric current density. This identity provides the most transparent route to the *calculation* of S. Suppose that an electric field E is applied to an isothermal homogeneous system. Then $E' = E$ and $\nabla T = 0$. Hence (18) gives $J = \sigma E$ and $Q = \pi\sigma E$ so that

$$S = \frac{1}{T\sigma} \cdot \frac{Q}{J} \tag{19}$$

where Q and J remain to be calculated. Of course, a calculation based directly on the normal measurement procedure yields the same result when the system has time-reversal symmetry. Nevertheless, it is usually conceptually more difficult because a temperature gradient must be introduced into the problem.

In this Section we consider the *diffusion* thermopower, S_d, i.e., we suppose that the phonons remain in thermal equilibrium and work out the behavior of the electrons on this basis. The assumption is obviously false: the drifting electrons are coupled to the phonons via the electron-phonon interaction. Consequently, there is an exchange of momentum between the electrons and the phonons which produces a heat flux in the phonon system, so that the phonons are *not* in thermal equilibrium. As a result: $S = S_d + S_g$ where S_g is the *phonon drag* contribution to the thermopower which we consider in Section 4 (Blatt 1968).

3.2. The Diffusion Thermopower Predicted by Boltzmann Equation for a Quasi-2D Electron Gas

In Section 2 we calculated the electrical conductivity of quasi-2D and quasi-1D electron gases at 0 K.

The diffusion thermopower vanishes at 0 K because it is impossible to set up a nonzero heat flux. To calculate S_d it is therefore necessary for us to calculate both J and Q at finite temperatures. Fortunately, both calculations are very easy and the results are well known for noninteracting electron gases subject to elastic scattering (Butcher 1973). Thus we have

$$J = \sigma E \tag{20a}$$

where

$$\sigma = \int d\varepsilon \;\; \sigma(\varepsilon)\, [-f_0(\varepsilon)] \tag{20b}$$

with $\sigma(\varepsilon)$ denoting the 0 K conductivity calculated in Section 2 and $f_0'(\varepsilon)$ denoting the energy derivate of the Fermi-Dirac function

$$f_0(\varepsilon) = [\exp(\varepsilon - \varepsilon_F)/k_B T) + 1]^{-1} \tag{21}$$

in which ε_F is the Fermi level. We note that when $T \rightarrow 0$, $-f_0(\varepsilon) \rightarrow \delta(\varepsilon - \varepsilon_F)$ so that (20) reproduces the result $\sigma = \sigma(\varepsilon_F)$ which we have already expressed in various forms in Section 2.

To proceed with the calculation of S_d we write (20) in the form

$$J = E\, \Phi\, d\varepsilon\, \sigma(\varepsilon)\, [-f_0(\varepsilon)] \tag{22}$$

which has a simple interpretation: the electric current density contributed by the electrons with energies between ε and $\varepsilon + d\varepsilon$ is

$$\delta J = E\, d\varepsilon\, \sigma(\varepsilon)\, [-f_0(\varepsilon)]. \tag{23}$$

The corresponding particle flux is $\delta J/(-e)$ and the corresponding heat flux is therefore

$$\delta Q = \frac{\varepsilon - \varepsilon_F}{-e} E \, d\varepsilon \, \sigma(\varepsilon) \, [-f_0'(\varepsilon)]. \tag{24}$$

By summing over all elements of the energy range we therefore find that

$$Q = -\frac{E}{e} \int d\varepsilon(\varepsilon - \varepsilon_F) \, \sigma(\varepsilon) \, [-f_0'(\varepsilon)] \tag{25}$$

hence (19) gives

$$S_d = -\frac{1}{eT\sigma} \cdot \frac{\int d\varepsilon(\varepsilon - \varepsilon_F)\sigma(\varepsilon)[-f_0'(\varepsilon)]}{\int d\varepsilon \;\; \sigma(\varepsilon)[-f_0'(\varepsilon)]}. \tag{25a}$$

The above argument is very puzzling. Fortunately the result, Equation (25), also follows from the Boltzmann transport equation when the electron-electron and electron-phonon interactions are both switched off (Butcher 1973). Moreover, and this is important for our consideration of the effect of level broadening in Section 4, (25) also follows from a many-body treatment of electron transport when these interactions are switched off (Mahan 1981).

The numerical evaluation of (25) in the cases of interest is elementary because $\sigma(\varepsilon)$ has already been calculated in Section 2 both with and without level broadening.

We consider here the zeroth order result obtained from the Boltzmann equation and we concentrate on the case when the electron gas is highly degenerate. The integrals (25) are evaluated for this case in most studies (Blatt 1969; Butcher 1973) on the assumption that $\sigma(\varepsilon)$ varies slowly in comparison to $f_0(\varepsilon)$. The result is:

$$S_d = \frac{\pi^2}{3} \frac{K_b^2 T}{e} \frac{\sigma'(\varepsilon_F)}{\sigma(\varepsilon_F)} \tag{26}$$

where the last factor is just the logarithmic derivative of $\sigma(\varepsilon)$ evaluated at the Fermi level.

Equation (26) remains valid in a quasi-2D system at low temperatures as long as $\sigma(\varepsilon)$ does not vary too fast with ε. A naive application of this equation using $\sigma(\varepsilon)$ as given in Figure 1 produces a negative contribution to

S_d (proportional to T), as ε_F moves *between* subband minima and a *positive*, δ-function-like, contribution to S_d (with a strength in the order of k_BT is no longer valid and a more careful evaluation of the integrals in (25) is necessary. When this is carried out (Cantrell and Butcher 1985b) we find that S_d is the sum of a negative contribution which is proportional to T and behaves in the way shown in Figure 5 when ε_F moves through a subband minimum and a positive contribution behaving in the way shown in Figure 6. This is the δ-function peak found in the elementary discussion. In reality it is thermally broadened to a width $-k_BT$ at temperature T. Since the strength (i.e., they are under the curve) is in the order of k_BT its *height* remains constant ($-2k_BT$ in $2/e$) as $T \to 0$. Consequently, S_d must change sign as $T \to 0$ when ε_F lies in the neighborhood of a subband minimum. This behavior has never been unambiguously observed, although Cantrell (1986) suggests that it may be present in some recent data of Davidson et al. (1986).

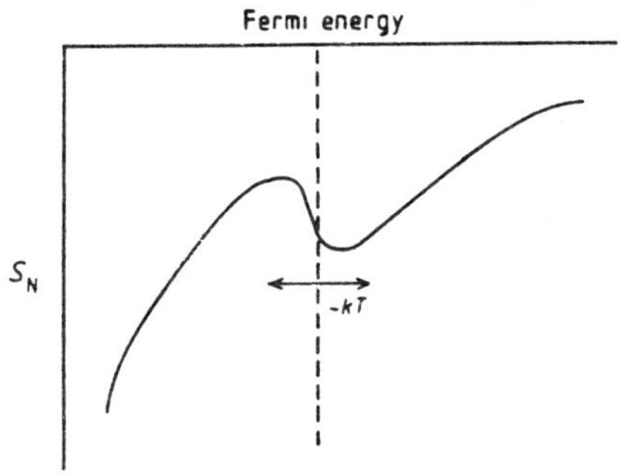

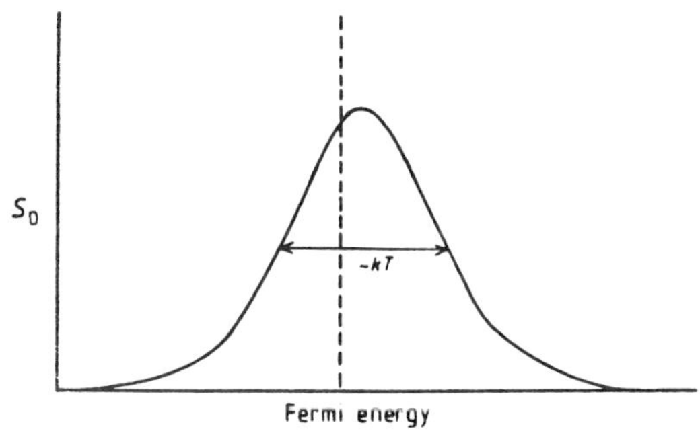

Fig. 5. Schematic plot of the negative contribution of S_N to the thermopower against Fermi energy in the neighborhood of a subband minimum indicated by the dashed line.

Fig. 6. Schematic plot of the positive contribution of S_D to the thermopower against Fermi energy in the neighborhood of a subband minimum indicated by the dashed line.

3.3. Results for a Quasi-2D Electron Gas

We pointed out in the remarks following (25) that the equation is valid beyond the Boltzmann transport regime. We may use it to calculate the effect of level broadening by substituting $\sigma(\varepsilon)$ for this case which has already been evaluated in Section 2.3. Numerical evaluation of the integrals yields the results presented in Figures 7-9 in which the subband minima are assumed to lie at 0, 5, 10, and 15 MeV. In Figure 7, which is drawn for T = 1 K, the case $(k_F l)_{2D}$ = 50 (full curve) represents the Boltzmann transport limit. It shows the very large effect of the positive quasi-δ-function peaks in the neighborhood of the subband minima. We see from the dashed curve ($(k_F l)_{2D}$ = 6) that the peaks remain at this scattering level but the sign changes are eliminated except near the second subband minimum. Figure 8 shows plots of S_d against T for $(k_F l)_{2D}$ = 6 when ε_F is near the second subband at 5 MeV. We find the normal metallic predicted by (26) when ε_F is well away from 5 MeV, and also when it is just above 5 MeV. The sign change which is expected from Figure 7 when $T \to 0$ appears for T - $\frac{1}{2}$ K when ε_F = 4.8 MeV. In Figure 9 we plot S_d against ε_F for T = 1 K and 6 K when $(k_F l)_{2D}$ = 2 which is a strong scattering case. The details are washed out but there is plenty of residual structure which should be observable.

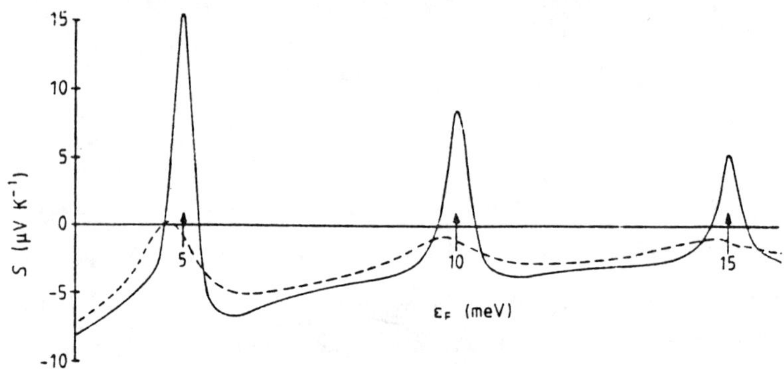

Fig. 7. A plot of S_d against Fermi energy. Full curve: $(k_F l)_{2D}$ = 50; broken curve $(k_F l)_{2D}$ = 6; both are for a temperature of 1 K.

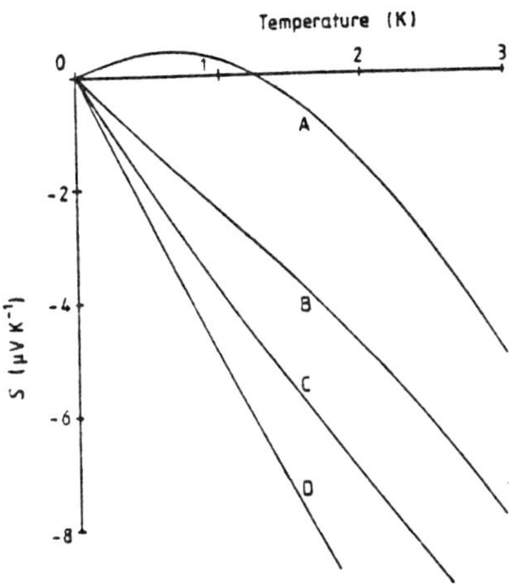

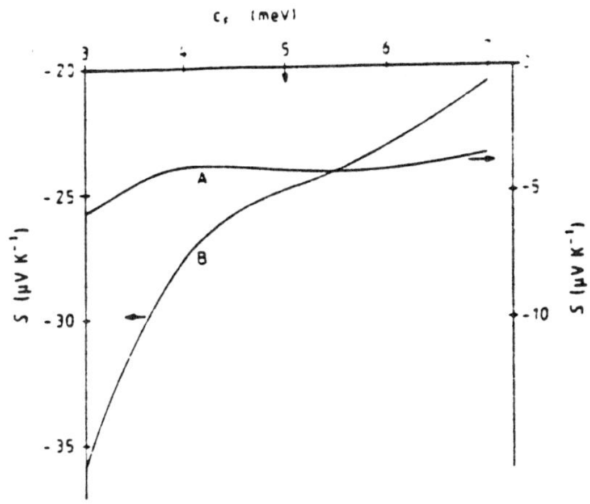

Fig. 8. S_d against temperature for $(k_F l)_{2D} = 6$ for values of ε_F (MeV) of 4.8 (a), 5.2 (B), 4 (C), and 6 (D), around a subband minimum at 5.0 MeV.

Fig. 9. S_d against Fermi energy for $(k_F l)_{2D} = 2$ for $T = 1$ K (A) and 6 K (B).

3.4. Results for Quasi-1D Electron Gas

The calculations made above for a quasi-2D electron gas may be extended to a quasi-1D electron gas by using the results for $\sigma(\varepsilon)$ for the case discussed in Section 2 (Kearney and Butcher 1987). We show typical results in Figure 10 for $(k_F l) = 30$ (i.e., the Boltzmann transport limit) and Figure 11 for $k_F l = 5$ (intermediate scattering). The positive peaks are clearly evident in both cases when ε_F is near the second and third subband minima at 5 and 10 MeV. They produce changes of sign of S_d as $T \to 0$ with ε_F just below either subband minimum.

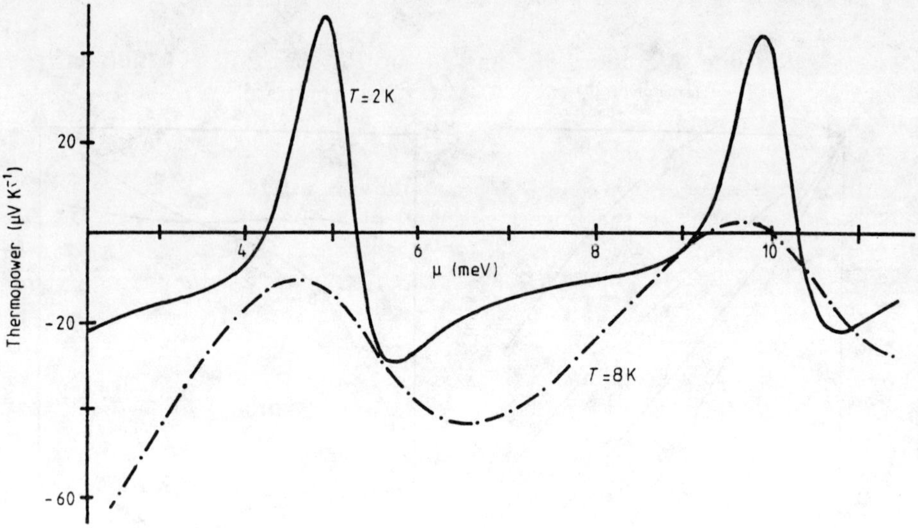

Fig. 10. Thermopower as a function of chemical potential for $(k_F l)_{1D} \simeq 30$. The second sub-band minimum is at 5 MeV, the third at 10 MeV.

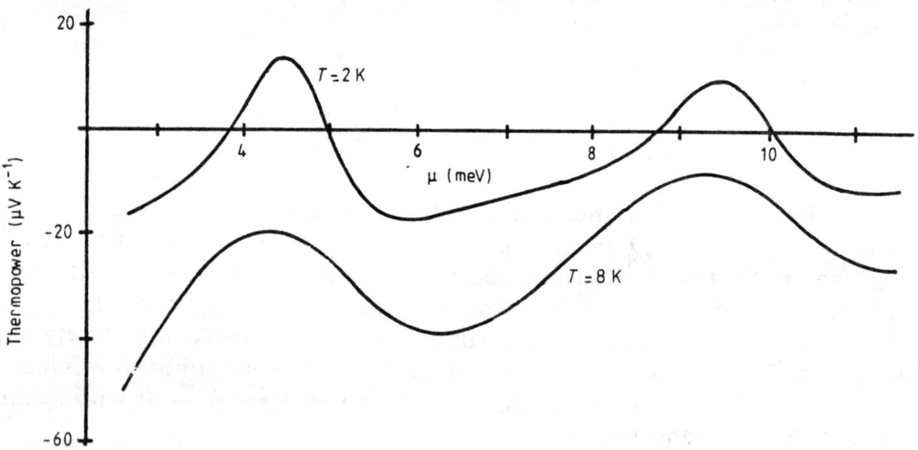

Fig. 11. The thermopower as a function of chemical potential for $(k_F l)_{1D} \simeq 5$.

4. Phonon Drag in a 2D Electron Gas Interacting With 3D Phonons

We outline the theory of Cantrell and Butcher (1986) which has been used to analyze data for a GaAs/AlGaAs heterojunction according to Fletcher et al. (1986) and data for a MOSFET according to Gallagher et al. (1986, 1987). Details of the theory will be published shortly in two papers (Cantrell and Butcher 1987a, b). We confine our attention to the case when the Fermi level lies in the lowest subband of a quasi-2D electron gas. The electrons are strictly 2D, while on the other hand, the phonons have a 3D nature. To simplify the calculation we treat them as strictly 3D and ignore the effect of reflections at any of the planar interfaces which are necessarily present in the systems studied experimentally.

The transport in this system is described by electron and phonon Boltzmann equations which are coupled by deformation potential electron-phonon scattering. These coupled equations are treated in a manner similar to that of Bailyn (1958). The equations are linearized and then solved to yield a single equation, of Boltzmann form, for the electron distribution function $f(k)$:

$$- \left.\frac{\partial f(k)}{\partial t}\right|_{\text{drift}} - U(k) = L(g(k)) \tag{27}$$

Here $g(k)$ is the perturbation to the equilibrium electron distribution function; $L(g(k))$ contains a number of terms, one of which describes the scattering of electrons off equilibrium phonons; and $U(k)$ is a new term produced by the temperature gradient. It describes the effect on the electrons of nonequilibrium phonons produced by the temperature gradient and is responsible for phonon drag.

Equation (27) is solved in the lifetime approximation with $U(k)$ treated as a quasi-drift term; i.e., we write

$$- \left.\frac{\partial f(k)}{\partial t}\right|_{\text{drift}} - U(k) = L(g)k = \frac{1}{\tau(k)} \frac{\partial f_0(k)}{\partial \varepsilon} g(k) \tag{28}$$

where $\tau(k)$ is the lifetime of the electron in state k. $\tau(k)$ can be calculated in the usual way. For simplicity we assume here that there are impurities in the conducting channel and that it is scattering off these that limits the electron lifetime. Rewriting the drift term in Equation (2) in the usual manner and setting $U(k) = H(k) \cdot \nabla T$, we find that

$$\left\{ \left[eE - (\varepsilon - \varepsilon_F)\nabla \ln T \right] \cdot \nu \frac{\partial f}{\partial \varepsilon} \right\} - H(k) \cdot \nabla T = \frac{1}{\tau(k)} \frac{\partial f_0(k)}{\partial \varepsilon} g(k) \tag{29}$$

which can easily be solved to obtain $g(k)$. The thermopower can then be found by calculating the electric current J and finding values of E and ∇T such that $J = 0$. We find that

$$E = (S_d + S_g)\nabla T \tag{30}$$

where S_d is the diffusion thermopower and S_g the phonon drag contribution. At temperature T this is given by

$$S_g = \frac{|e|}{AT\sigma}\sum \tau(k)H(k)\cdot\nu(k) \tag{31}$$

where A is the area of the two-dimensional system, σ its electrical conductivity, and $\nu(k)$ the velocity of electrons with 2D wave-vector k. In detail the formula for S_g, which is similar to that given by Bailyn (1967), is

$$\begin{aligned} &\frac{|e|}{AT\sigma}\sum_{q,\,q_z}^{|q|}\sum_{k,\,k'} A(q,\,q_z)\,\frac{n_{0q,q_z}}{k_BT}\quad f_0(k')(1 - f_0(k')\delta(E-)\delta(k-)\,\times \\ &\times\,(\nu(k)\tau(k) - \nu(k')\tau(k'))\cdot\nabla\omega\,\frac{\partial n_0}{\partial\gamma}\,\frac{\gamma}{F(q,\,q_z)} \end{aligned} \tag{32}$$

where

$$A(q,\,q_z) = \frac{\pi\varepsilon_1^2(q^2 + q_z^2}{\rho V\omega_{q,q_z}}\,|\int\phi^2(z)e^{iq_z z}dz|^2 \tag{33}$$

describes the (assumed) deformation potential scattering of the electrons by the phonons. In these equations ε_1 is the deformation potential, ρ is the material density, V is the sample volume and $\pi_{q,\,q_z}$ is the angular frequency of the phonon mode with wave-vector $(q,\,q_z)$. In labeling the phonon mode, q is the component of phonon wave-vector parallel to the electron layer and q_z is the perpendicular component. The squared matrix element contains information about the shape and width of the electron sheet, $\phi(z)$ being the wavefunction of the ground subband. We have also used the notation

$$\delta(E-) = \delta(\varepsilon_{k'} - \hbar\omega_{q,q_z} - \varepsilon_k) \tag{34}$$

$$\delta(k-) = \delta(k' - k - q) \tag{35}$$

$$n_{0q,q_z}\quad\nabla k_BT. \tag{37}$$

$F(q,\,q_z)$ is the sum of the scattering rates for all events that a phonon of wave-vector $(q,\,q_z)$ may take part in, i.e., the sum of the phonon-phonon and phonon-electron scattering rates. It can be shown that the largest term is the phonon-phonon rate which, when such scattering is treated in a lifetime of approximation, is given by (Bailyn 1958):

$$-(\partial n_{0q,q_z})/\partial\gamma)(1/k_BT)/\tau_p(q,\ q_z) \tag{38}$$

with $\tau_p(q,\ q_z)$ denoting the phonon-limited phonon lifetime, i.e., the time between phonon-phonon scattering events. To show that (38) is the largest term it is necessary to include the lifetime broadening of the electronic states in a phenomenological manner as the bare phonon-electron scattering rate is divergent in two places. In (32) the sum over k and k' is limited to all pairs of these vectors that have a difference of q. This is indicated by writing $|q|$ over the summation sign. Formula (32) can be evaluated at low temperatures to yield

$$S_g = -B \int_{-q_D}^{q_D} dq_z| \int \phi^2(z) e^{iq_z z} dz|^2 \times$$
$$\times \int \frac{}{\left[(\hbar^2 k_F q/m^*)^2 - \left\{\hbar\omega_{q,q_z} - (\hbar^2 q^2/2m^*)\right\}^2\right]^{1/2}(e^\gamma - 1)^2} dq \tag{39}$$

where

$$B = (3\hbar m^* \varepsilon_1^2 \nu^2/8\pi^3 k_B T^2 n_0 |e|\rho) \tag{40}$$

where n_0 is the real density of electrons. We have assumed Debye phonons of velocity m.

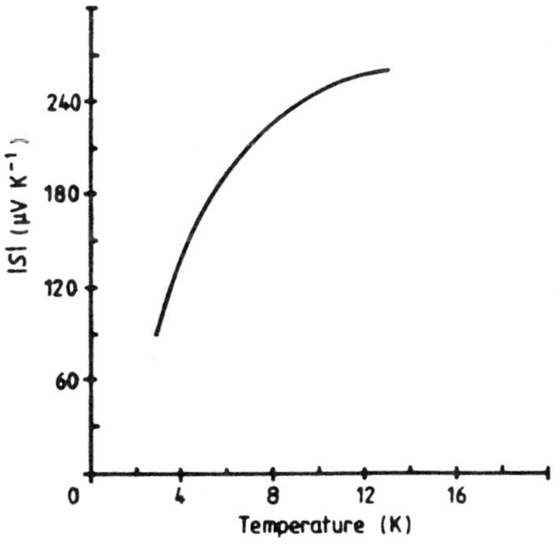

Fig. 12. A plot of S_g against T from results of Cantrell and Butcher (1986). S_g (at 5 K) is fitted to the data of Fletcher et al. (1986).

In the data of Fletcher et al. (1985) there is a peak in the drag at ~13 K. This peak is due to the behavior of $\tau_p(q, q_z)$ as a function of temperature. Above the peak temperature τ_p will be dictated by phonon-phonon collisions: these become much less frequent as the temperature is lowered and hence τ_p rises quickly giving the increase in S_g on the high-temperature side. Eventually τ_p becomes limited by phonon collisions with the walls of the sample and it stays constant as temperature is lowered. In this regime τ_p can be taken from the integral in (38) and combined with the speed of sound ν, to yield the phonon mean free path L, which is now the sample dimension. When τ_p is constant S_g quickly falls due to the Bose factors in the integral. Once the temperature is low enough such that there are very few phonons of this wave-vector, the drag is small.

Figure 12 shows a comparison of the temperature dependence predicted by (39) with the experimental data of Fletcher et al. (1985) in the low-temperature regime in which boundary scattering of the phonons dominates. Only one curve is shown because the temperature dependence observed is correctly predicted by the theory.

It is only necessary to scale up the theoretical values by a factor of 2 to get good numerical agreement for $T < 12$ K. An error factor of 2 is not very surprising in view of the many approximations made in deriving (39). Work is in progress to remove them, and to investigate quantum size effects in S_g and the influence of level broadening. Gallagher et al. (1986; 1987) describe the more extensive comparison between theory and experiment which has been made for an Si MOSFET.

5. Conclusion

We have surveyed some of the problems currently being considered in the theory of electron transport for quasi-2D and quasi-1D electron gases. The ability to fabricate systems containing low-dimensional electron gases has been a tremendous stimulus to electron transport physics. Once made, the systems can be moved to another laboratory where the basic requirements for their experimental study are a voltmeter and an ammeter. On the theoretical side, a great many problems remain. To give just two examples: we have outlined a very elementary treatment of phonon drag, which requires many extensions; we have ignored altogether the important topic of the screening of electron-phonon interactions.

Interest in low-dimensional systems (LDS) is not confined to electron transport physics. They also have interesting magnetic, dielectric, and vibrational properties. LDS represent a new class of materials which are ripe for investigation and which exhibit many properties not found in naturally occurring solids. The study of their physics and applications provides us with a fascinating new challenge.

References

1. Bailyn, *Phys. Rev.*, vol. 112, p. 1587, 1985.
2. Blatt, F. J., *Physics of Electronic Conduction in Solids*, New York: McGraw-Hill, 1968.
3. Butcher, P. N., *Electronics in Crystalline Solids*, ed. by A. Salam, Vienna, IAEA, 1973.
4. Cantrell, D. G., *J. Phys. C*, vol. 19, p. L817, 1986.
5. Cantrell, D. G., and Butcher, P. N., *J. Phys. C*, vol. 18, p. 5111, 1985a.
6. Cantrell, D. G., and Butcher, P. N., *J. Phys. C*, vol. 18, p. L587, 1985b.
7. Cantrell, D. G., and Butcher, P. N., *J. Phys. C*, vol. 19, p. L429, 1986.
8. Cantrell, D. G., and Butcher, P. N., To be published in *J. Phys. C*, 1987b.
9. Cantrell, D. G., and Butcher, P. N., To be published in *J. Phys. C*, 1987b.
10. Davidson, J. S. et al., *Phys. Rev. B*, vol. 33, p. 8238, 1986.
11. Fletcher, R. et al., *Phys. Rev. B*, vol. 33, p. 7122, 1986.
12. Gallagher, B. et al., To be published in *J. Phys. c*, 1987.
13. Kearney, M. J., Butcher, P. N., *J. Phys. C*, vol. 19, p. 542, 1986.
14. Kearney, M. J., Butcher, P. N., *J. Phys. C*, vol. 20, p. 47, 1987.
15. Maines, P., private communication, 1987.
16. Mahan, G. D., *Many Particle Physics*, New York: Plenum, 1981.
17. Ogrin, Yu. et al., *JETP Letts.*, vol. 3, p. 71, 1986.
18. Poole, D. A. et al., *J. Phys. C*, vol. 14, p. L995, 1981.
19. Sernelius, et al., *J. Phys. C*, vol. 18, p. 225, 1985.
20. Siggia, E. D., Kwok, P. C., *Phys. Rev. B*, vol. 2, p. 1024, 1970.
21. Zavaritskii, N. V., Kron, Z. D., *JETP Letts.*, vol. 38, p. 97, 1983.

THE ELECTRICAL TRANSPORT PROPERTIES OF QUASI TWO-DIMENSIONAL ELECTRONS

B. Gallagher
Department of Physics
University of Nottingham
Nottingham, NG72RD, UK

Abstract: **Recent developments in the growth of epitaxial semiconductor layers and in lithographic techniques have led to a new range of structures in which the conduction electrons have a quasi two- or one-dimensional nature. In this review a brief survey of the electrical transport properties of quasi two-dimensional electrons is given. First, the fabrication of devices is discussed. Then starting from the most basic properties of two-dimensional electrons an introduction is given to the most dramatic manifestation of their two-dimensional nature, the quantum Hall effect. Finally, one of my own particular research interests, the thermopower of two-dimensional electrons, is briefly considered.**

1. Introduction

In any physical system electrons are obviously three-dimensional in the sense that electron density must have a finite spatial extent in all three dimensions. However, if electrons are confined in one of these dimensions, z, for example, by a potential well of width comparable with their wavelength solution of the Schrödinger equation gives discrete energy levels for this dimension. We then have a situation where an external applied field produces normal band conduction in the x-y plane but where motion in the z-direction only occurs by transition between these discrete energy levels. If the energy separation between these levels is large compared with the thermal energy kT, and the electron density n_s is sufficiently low, all the electrons will be in the lowest level and no motion in the z-direction is possible. The electrons are then said to be two-dimensional in the sense that their motion is strictly confined to the x-y plane. Similarly, confinement in two dimensions can lead to one-dimensional behavior. Such behavior occurs at low temperatures in semiconductor material for potential wells with a typical width of ~50 Å.

Electron confinement on these length scales occurs at the Si/SiO_2 interface of metal-oxide-silicon field effect transistors (MOSFETs) and at $GaAs/Al_xGa_{1-x}As$ interfaces. The most important devices in modern microelectronics depend upon the properties of electrons at such interfaces and thus upon the two-dimensional electron gas (2DEG). Therefore, there has been an extremely fruitful interplay between the technology of microelectronics and the fundamental physics of low-dimensional structures. Improvements in our understanding of the quantum properties of such structures guiding the development of devices, and the refinement of techniques for their fabrication make available specimens upon which to test new theories and ideas.

The study of low-dimensional structures has been the major growth area within semiconductor physics for the last five or six years, and it is not my intention to try to survey this vast field. I will concentrate on some basic aspects of low-dimensional electrical transport. The emphasis will be on experimental results and a simple exposition of the underlying physics. The major problem facing an experimenter in this field is always the availability and quality of specimens. An experimenter must have some understanding of the technology by which specimens are produced since it is the technology which usually sets the limitations on physics. I will thus start with an introduction to some basic device structures.

2. The Si MOSFET

During the 1970s the Si MOSFET was by far the most important system used in the study of the two-dimensional electron gas (2DEG). Work up to 1981 is reviewed in a definitive study by Ando et al. (1982), while Pepper (1985) presents a very useful introduction to the subject and more recent results.

Figure 1 illustrates a typical Si MOSFET. We can view the system as essentially a capacitor in which the metal gate and the semiconductor are the plates and the oxide is the dielectric. If the semiconductor is *p*-type and we apply a positive bias to the gate, holes (the majority carrier) are repelled from the Si/SiO_2 interface leaving charged acceptors in the depletion region. Eventually as the bias is increased the conduction band falls below the Fermi level so that a layer of electrons forms at the surface. This is known as an inversion layer since minority carriers (electrons) are now majority carriers. For *n*-type substrates and negative gate voltages hole inversion layers can be formed. Once the inversion layer starts to form it screens the interior of the semiconductor so that almost all further charge is induced in the inversion layer. The electron density is then simply proportional to the gate voltage and the depletion layer width remains essentially constant. If, as is shown, contact is made to the inversion layer by a heavily *n*-doped (n^+) source and drain contacts, the current can be controlled by the gate bias. The width of the inversion layer so formed is typically 20 Å for the Si (1,0,0) surface. This is the surface of most interest technically and scientifically, since electrons in the inversion layer have the highest mobility on this surface. We thus expect the electrons to be two-dimensional at low temperatures. The great attraction of the Si MOSFET for the study of 2D behavior is that using the gate voltage we can directly vary the carrier density over more than two orders of magnitude. The upper limit of $\sim 10^{17}$ m^{-2} is set by the breakdown field of the oxide (-10^7 V/cm), the lower limit of $-5 \cdot 10^{14}$ m^{-2} due to the disorder at the interface localizing the electrons (this will be discussed later).

It can be seen from Figure 2a that the electrons are in a potential well which is roughly triangular. The energy levels in the *z*-direction are determined by the band bending and the modification to the potential produced by the electrons themselves. Stern and Howard (1967) calculated these levels and the corresponding charge distribution by self-consistently solving the Poisson and Schrödinger equations. An example of the charge distribution obtained is shown in Figure 2b. Note that nearly all the charge is in a layer -20 Å thick and that the electrons stand away from the interface. This second result is easily understood since the energy barrier at the interface is -3 eV which is much larger than the electron energies. The interface therefore presents an almost infinite potential and the wavefunctions have very

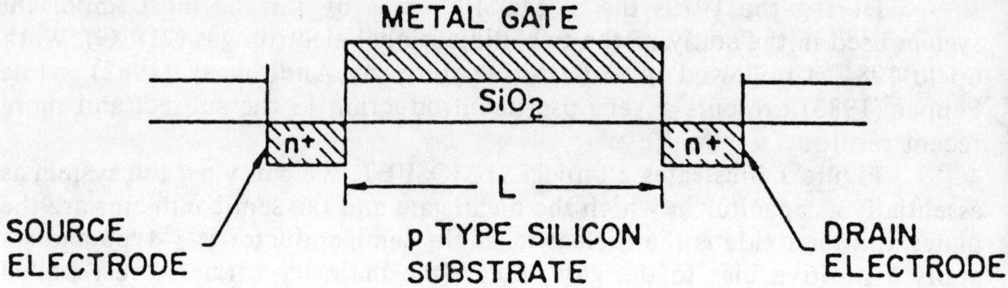

Fig. 1. A cross-section of an n-channel Si MOSFET. The n^+ contacts are made by diffusion or by ion implantation into the p-type substrate. The current between them is controlled by the gate voltage.

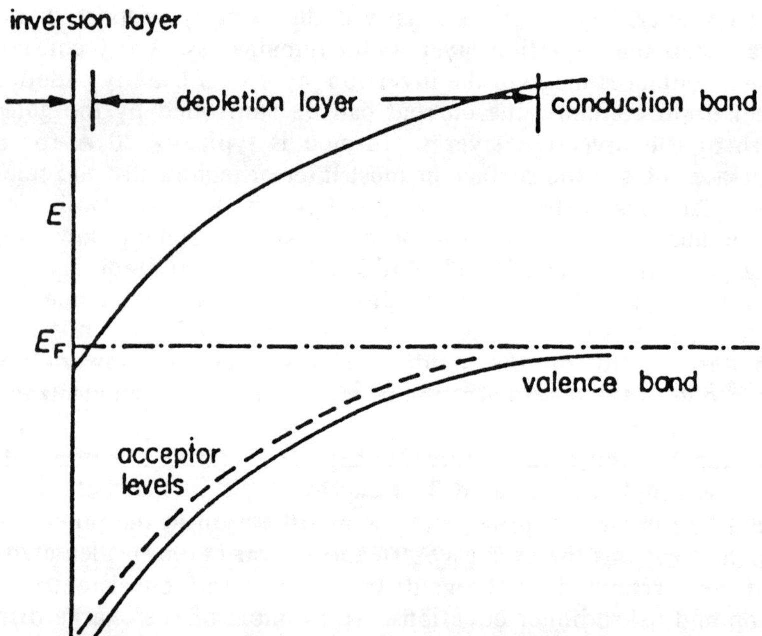

Fig. 2. a) Band bending at the Si surface leading to the formation of an inversion layer.

little penetration into the oxide. Thus the charge density is close to zero at the interface. The solutions for the electron states in the plane of the interface are free electron-like with an effective mass m^* associated with the conduction band minima. To a good approximation the energy states can be written as

$$E = E_n(m_z^*) + \frac{\hbar^2 k_x^2}{2m_x^*} + \frac{\hbar^2 k_y^2}{2m_y^*}. \tag{1}$$

In silicon we have the complication of an anisotropic effect mass, m^* and a six-fold valley degeneracy. however, for the (1,0,0) surface the degeneracy is broken and the lowest subband has $m_z^* = 0.92$, $m_x^* = m_y^* = 0.19$ and a two-fold valley degeneracy. The separation between the lowest subband energy and the first excited state is ~25 MeV for normal substrate doping levels. So below ~30 K the electrons will be strictly two-dimensional.

For the experimenter the Si MOSFET is an extremely attractive system. It has great flexibility due to its ability to control electron density. The inversion layer thickness may also be varied to an extent by application of a substrate bias. It also relies on a very well developed and sophisticated technology which provides ohmic contacts that work at low temperatures, gates which do not leak, and good sample-to-sample reproducibility. Working with any other semiconductor material is often more problematic in these respects. However the Si MOSFET has the basic problem of rather low mobilities. Scattering from charge trapped in the SiO_2 and from surface roughness leads to typical mobilities of ~8000 $cm^2\ V^{-1}\ s^{-1}$ at 4.2 K. In a few outstanding specimens values of ~25,000 $cm^2\ V^{-1}\ s^{-1}$ have been obtained (see Smith and Stiles 1986, for example) but this is still much less than the values obtainable for GaAs/AlGaAs heterostructures.

3. GaAs/AlGaAs Heterostructures and Quantum Wells

It has long been recognized that the semiconductor compound GaAs has considerable advantages over Si. It has a much lower effective electron mass and hence much higher room temperature mobility; ~10,000 $cm^2\ V^{-1}\ s^{-1}$ compared with ~1300 $cm^2\ V^{-1}\ s^{-1}$. So higher frequency responses, shorter switching times, and lower power dissipation can be achieved in GaAs devices. It is also a direct bandgap semiconductor so that, in principle, conversion of electrical energy into light and ice versa is possible. All optoelectronics are based on GaAs and similar III-V semiconductors. Many of these optoelectronic devices involve structures in which th4e electrons and holes are low-dimensional. A useful review of this field is given by Narayanamurti (1984).

It might be assumed that we could make a structure similar to Si MOSFET out of GaAs. This turns out not to be so. The Si MOSFET relies on the very small number of surface states produced at the Si/SiO_2 interface. GaAs does not have a stable native oxide and producing an interface of similar quality is far from easy. It was realized when this problem was first considered that since $Al_xGa_{1-x}As$ was very well lattice matched to GaAs over the whole composition range we could, in principle, use the inversion layer formed at this interface as the conducting canal of an FET. In practice however, growing such interfaces proved extremely difficult. Liquid or vapor phase epitaxy techniques, in which the semiconductor is grown by deposition from the liquid or vapor medium, are not well suited to producing abrupt interfaces. In recent years metal-organic chemical vapor deposition (MOCVD) has been used to produce a surprisingly good interface (see Mullin et al. 1984) but it was with the advent of molecular beam epitaxy (MBE) that the great advances were made.

MBE really only involves controlled evaporation onto a substrate of the compound semiconductor constituents. However the extremely clean ultra-high vacuum environment keeps unintentional doping by impurities to an absolute minimum. Furthermore, the rate of growth of the semiconductor surface can be very closely controlled. Slow, layer by atomic layer, growth allows accurate independent control of thickness, doping, and composition. By this technique we can make interfaces which are atomically abrupt and which are in fact far superior in quality to the Si/SiO_2 interface. However, the mobility of carriers in the GaAs/AlGaAs inversion layer were still found to be rather low, until modulation doping was introduced in 1978. The simple idea behind this is to spatially separate the conduction electrons from their parent donor impurity atoms thereby greatly reducing the influence of ionized and neutral impurity scattering. That is, rather than uniform doping, we only dope in regions away from the conducting channels. When this is done mobility can be greatly enhanced, particularly at helium temperatures where impurity scattering normally completely dominates (see Stormer (1983) for a useful review). Using modulation doping new records for both electron and hole mobilities are constantly being set. The extremely rapid progress is illustrated in Figure 3. The current record is ~3,000,000 $cm^2\ V^{-1}\ s^{-1}$ which corresponds to mean free paths on the order of 20 μm. This opens up the possibility of a whole range of new quantum effects because the electrons can retain their phase over distances corresponding to the size of the devices (Stormer 1983). The possibility of ballistic transistors with very high speeds also exists. This technology has already led to the production of "modulation doped field effect transistors" (MODFETs) or "high electron mobility transistors" (HEMTs), as they are called, with very high speeds. At 77 K, where the modulation doping advantage is significant,

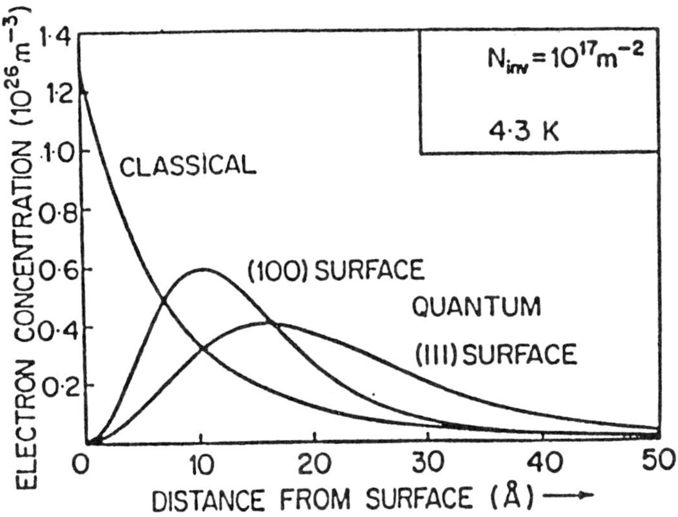

Fig. 3a. The calculated charge distribution within the inversion layer.

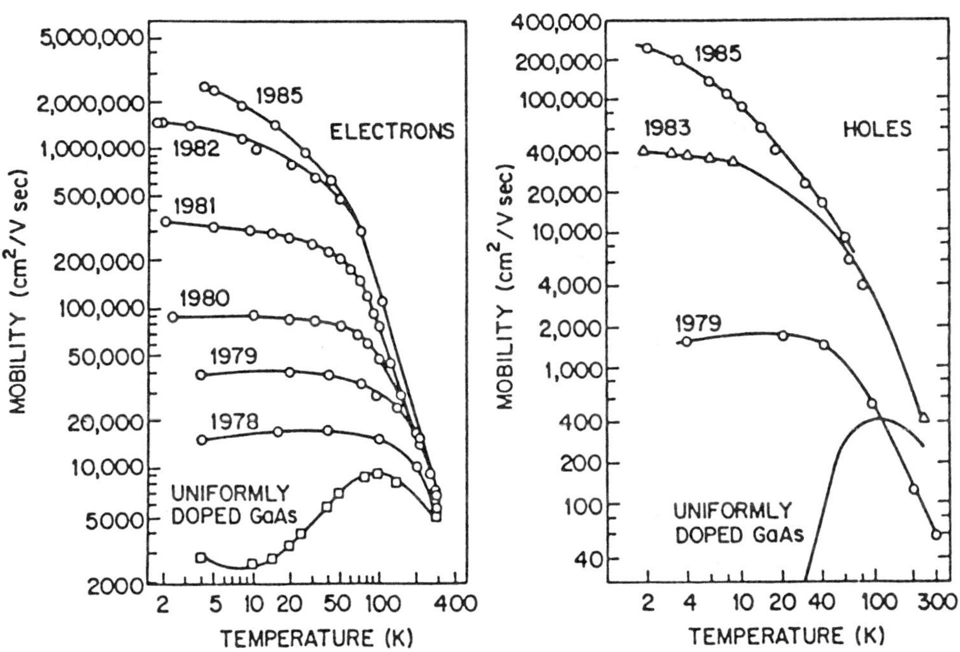

Fig. 3b. Highest reported (a) electron and (b) hole mobilities in GaAs/AlGaAs heterostructures.

switching times as short as 6 ps have recently been observed (Shah et al. 1986).

At an interface between two semiconductors we must have band bending because of the mismatch in Fermi energies and bandgaps. For $Al_xG_{1-x}As$ the bandgap is given by

$$Eg = 1.42 + 1.25\,x \quad \text{for } x < 0.45.$$

About 85% of the discontinuity is believed to be taken up in the conduction band giving an offset of ~x eV. Thus, for the commonly used GaAs/$Al_{0.3}Ga_{0.7}As$ interface the offset is ~0.3 eV. At a single heterojunction (Figure 4) the carrier density at the GaAs interface is determined by the level of doping in the AlGaAs layer. The GaAs is left nominally undoped since we wish the electrons to be in a region free of impurities. The width of the confining well must again be found self-consistently for Poisson's and Schrödinger equations. It is found to be of the same order as the electron wavelength and so we expect two-dimensional behavior.

In this arrangement the conduction electrons will still be scattered by impurities in the AlGaAs since the electron wavefunctions will penetrate a relatively long way because of the small energy barrier. This effect can be greatly reduced by the introduction of a thin undoped spacer region of AlGaAs at the interface. In MBE this is simply achieved by switching on the Si molecular beam after the Al. If however, the spacer layer is too large, the number of electrons in the confining potential will drop, since it is the donors in the AlGaAs which are holding the Fermi level at a high energy with respect to the conduction band edge. This leads to a drop in the Fermi velocity and to a reduction of screening. These factors then tend to reduce mobility. An optimum thickness is found to be ~150-200 Å. It should also be noted that mobility can be greatly enhanced at low temperatures by illumination. This increases carrier concentration by activation from deep traps. All the records for high mobilities have been achieved under illumination.

The GaAs/AlGaAs interface can be turned into an FET by applying a gate to the AlGaAs layer as shown in Figure 5. Evaporation of Al or a Ti-Au alloy onto the AlGaAs creates a Schottky barrier which is pinned at -0.8 eV by surface states. A depletion region is now formed beneath the gate. Application of bias to the gate increases or decreases the width of the depletion layer and thus the electron density in the potential well at the interface. In practice, however, the electron densities can only be changed by a factor of approximately five.

The GaAs/AlGaAs heterostructure is extremely important because of its high mobility which, for example, was crucial to the discovery of the fractional quantum Hall effect. It is however very difficult to obtain speci-

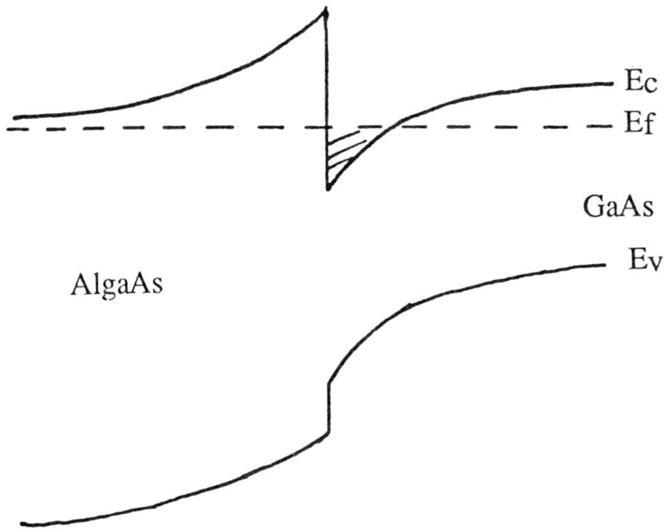

Fig. 4. Band bending at a GaAs/$Al_xGa_{1-x}As$ interface.

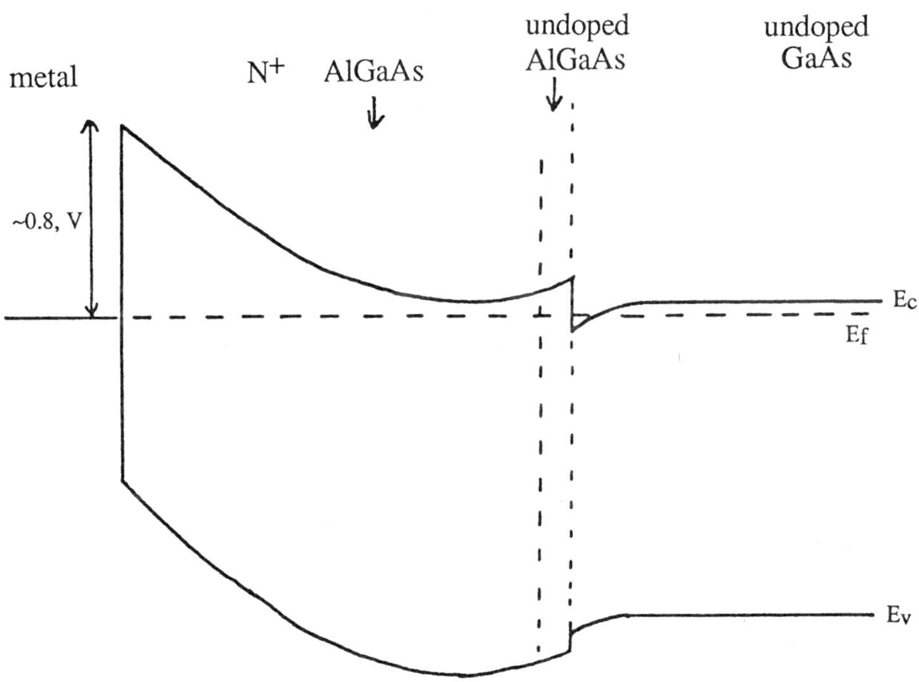

Fig. 5. Band diagram of a GaAs/AlGaAs HEMT.

mens unless one has a close collaboration with a group doing MBE growth. Even then the limitations of GaAs technology can lead to serious problems with reproducibility. A seemingly simple problem like putting a good low temperature ohmic contact onto GaAs can turn out to be very difficult. We expect that these problems will be solved rapidly as the commercial production of HEMTs gets underway.

4. Other Structures

A number of other important structures which MBE growth and high resolution lithography have now made available should be noted.

First, of course, we need not restrict ourselves to GaAs/AlGaAs. All the III-V compound semiconductors may be alloyed into one another leading to the possibility of growing different materials with the same lattice constants but different energy gaps. This is extremely important, for example, if we wish to match the bandgap to a particular wavelength of light.

It is also possible to make quantum wells by growing a thin (typically 20-200 Å) layer of narrower gap material between thick layers of wider gap material. As shown in Figure 6 we have then constructed the equivalent of the quantum mechanical particle in a box. The allowed energies in the z-direction will then be determined by the well width chosen. Attempts to produce high mobility modulation doped quantum wells have not, however, been very successful. If the layer growth sequence is repeated several times then we obtain a multiple quantum well structure (Figure 6). WE may then have overlap between the wavefunctions in adjacent wells if the layers are sufficiently thin and narrow bands rather than discrete energy levels. This is just like a one-dimensional Kroenig-Penney potential and the structure may be said to be a superlattice. this gives us the possibility of growing structures with a specified band structure. Capasso (1984) gives an introduction to the many periodic and aperiodic structures which can be produced.

Conduction electrons can be confined in more than one dimension by the use of electron beam lithography. This technique can now define structure as small as ~100 Å (Broers, 1986). There are two main approaches to this. Lithography may be used to define masks which are then used to physically restrict the dimensions by etching away the conducting material. Alternately such masks can be used to lay down metal gates which then restrict the motion of the electrons by the imposition of electrical field. Examples of these types of structures are shown in Figure 7.

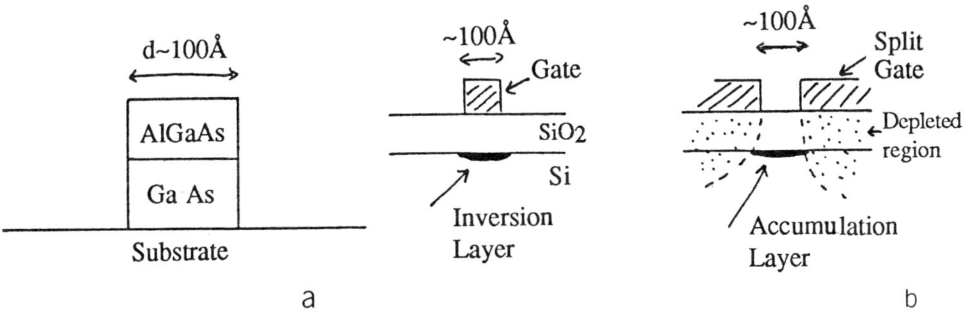

Fig. 6. Band diagrams for (a) single and (b) multiple quantum well structures.

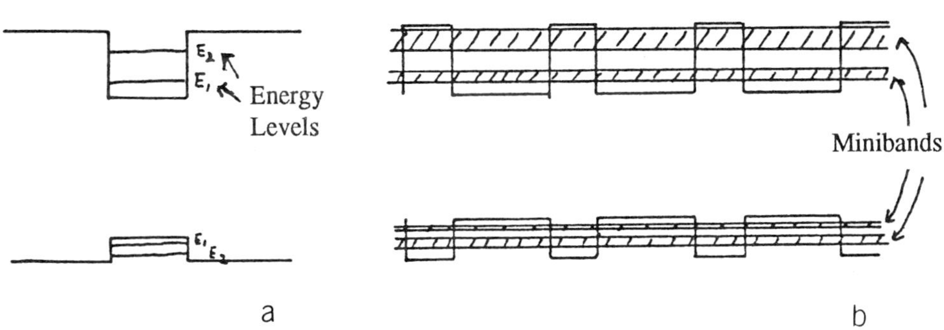

Fig. 7. Confinement of a 2 DEG to produce quasi-1D conduction by (a) direct etching and (b) a confining electrical field.

5. The Density of States in 2D

Let us consider two-dimensional free electrons. As noted earlier if the confinement is in the z-direction the allowed energy states are

$$E = E_n + \frac{\hbar^2}{2m}(k_x^2 + k_y^2). \tag{3}$$

Thus we have a series of electric subbands as is shown in Figure 8. In k-space the spherical constant energy surfaces of the 3D case are now replaced by constant energy circles. For one subband, if we consider the electrons to be confined to a square of side L, and apply periodic boundary conditions, the values of k_x and k_y are limited to

$$k_x = n_x \cdot \frac{2\pi}{L}; \qquad k_y = n_y \cdot \frac{2\pi}{L} \tag{4}$$

where n_x and n_y are integers. Allowing for the two spin states, each translational state then occupies an area in k-space of $2\pi^2/A$ where A is the area of the sample. So the number of states with wavevectors between k and $k + dk$ is:

$$\left(\frac{A}{2\pi^2}\right) \cdot 2\rho k \, dk = \frac{Am}{\pi\hbar^2} \, dE \tag{5}$$

where dE is the energy spread corresponding to dK. Therefore in 2D the density of states is a constant:

$$N(E) = \frac{m}{\pi h^2} \text{ per unit area.} \tag{6}$$

Each of the subbands will contribute the same constant density of states. The contribution from each beginning at the subband energy E_n is illustrated in Figure 8. In a real semiconductor this will still be correct if we introduce the effective mass m^* and allow for additional valley degeneracy. Scattering will, however, tend to smear out the sharp steps.

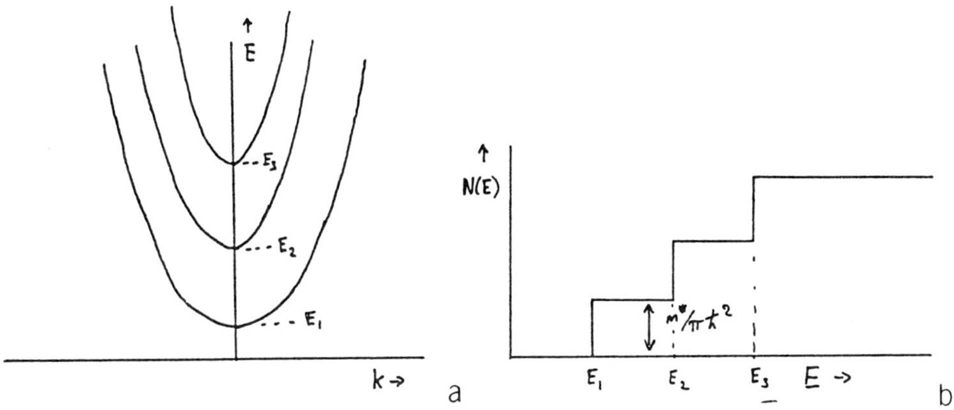

Fig. 8. Illustrations of (a) 2D electric subbands and (b) the 2D density of states.

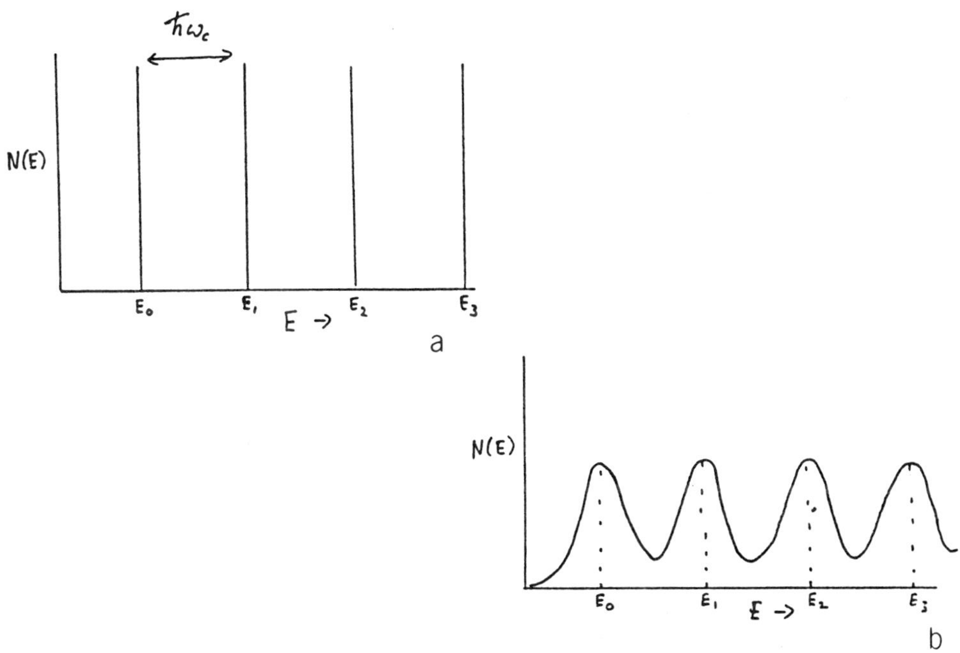

Fig. 9. Illustration of the density of states in a normal magnetic field in (a) the absence of scattering, (b) the presence of scattering.

6. The Effect of a Magnetic Field Perpendicular to 2DEG

Consider the electron motion in the two-dimensional layer with a magnetic field B normal to the surface. Classically the electrons move in orbits of radius r, with a cyclotron frequency ω_c, where

$$r = \frac{mv}{Be} \quad \text{and} \quad \omega_c = \frac{Be}{m} \tag{7}$$

To treat the same problem quantum mechanically we must choose a particular form for the magnetic vector potential A. Since the requirement is that B = curl A we can choose any of a number of forms, or gauges each of which will lead to states with the same energies but with different eigenstates. Choosing the asymmetric, or Landau, gauge $A = B(0,x,0)$, the hamiltonian is

$$H = \frac{(p + eA)^2}{2m} = \frac{p_x}{2m} + \frac{1}{2m}(p_x + eBx)^2. \tag{8}$$

Which can be written as

$$H = \frac{p_x{}^2}{2m} + \frac{1}{2}m\,\omega_c^2(x + x_0)^2 \quad \text{where } x_0 = p_y/m\omega_c. \tag{9}$$

This is the hamiltonian for a harmonic oscillator centered at x. The allowed energies are therefore

$$E = (1 + \tfrac{1}{2})\hbar\omega_c \quad \text{for } i = 0,\ 1,\ 2... \tag{10}$$

The "confinement" in the x-y plane imposed by the magnetic field therefore leads to discrete energy levels and the 2D electrons are now in totally quantized states. The quasicontinuum of transitional energy levels in the absence of the field now condenses into a set of discrete energy levels as is illustrated in Figure 9. These are called Landau levels and an ideal periodic system are delta-functions. In any real system we will have some disorder and scattering of electrons. Then the Landau levels will be broadened and for a relaxation time τ will have a width, $\Gamma \sim \hbar/\tau$. If this width is larger than the level separation, $\hbar\omega_c$, the levels will be blurred together. This will be the case for small B and thus small ω_c. However, for $\hbar\omega_c \gg \hbar/\tau$, i.e., $\omega_c\tau \gg 1$, the Landau levels will be narrow and well separated. Classically this corresponds to the electrons executing many orbits before they are scattered. At finite temperature we will also need $\hbar\omega_c \gg kT$ to obtain well resolved levels. The number of electrons occupying a Landau level, i.e., its degeneracy, can be easily obtained since it is only the number of states which existed in an energy range $\hbar\omega_c$ in the absence of the B field. In two dimensions this is

$$N(E) \cdot \hbar\omega_c = \frac{m}{\pi h^2} \cdot \hbar\omega_c = \frac{2Be}{h}. \tag{11}$$

That is, each Landau level has exactly the same degeneracy.

We have, however, missed the fact that in a magnetic field spin-up and spin-down electrons will have energies which differ by the factor $g^*\mu_B B$ where g^* is the effective Lande factor and μ_B is the Bohr magneton. Including this we have

$$E = (i + \tfrac{1}{2})\hbar\omega_c \pm \tfrac{1}{2}g^* \mu_B B. \tag{12}$$

Each Landau level will therefore be split into two, each with a degeneracy of Be/h. Again, whether or not these are resolved will depend on τ and thus mobility.

It is important to note that complete separation into Landau levels only occurs in 2D. In 3D the motion in the z-direction is not quantized and the density of states is still a quasicontinuum.

One final important point is that in 2D resistances and resistivities have exactly the same values and units. They are measured in units of per square and depend on specimen shape, but not on size.

7. The Shubnikov-de Hass Effect

If we measure the conductivity of the 2DEG as a function of B we find strong oscillatory behavior. This is because, as the magnetic field increases, the separation between the Landau levels is increased and successive levels pass through the Fermi level. Such Shubnikov-de Hass oscillations have been found in a number of 3D systems but the observed changes are only a few percent. In 2D very large oscillations can be observed. In a system like the Si MOSFET where we can vary the electron density and thus E_F we can also observe such oscillation by keeping the magnetic field fixed and sweeping E_F through the Landau levels. Examples of such measurements are shown in Figures 10 and 11. We see that the structure can indeed be very dramatic. The results for the Si MOSFET show the effects of spin splitting of the landau levels and also valley splitting which we have not considered. In the case of GaAs/AlGaAs heterostructure the effective mass is smaller and therefore the splitting of the Landau levels is larger. Also higher mobility leads to much narrower levels and much sharper oscillations. A key point

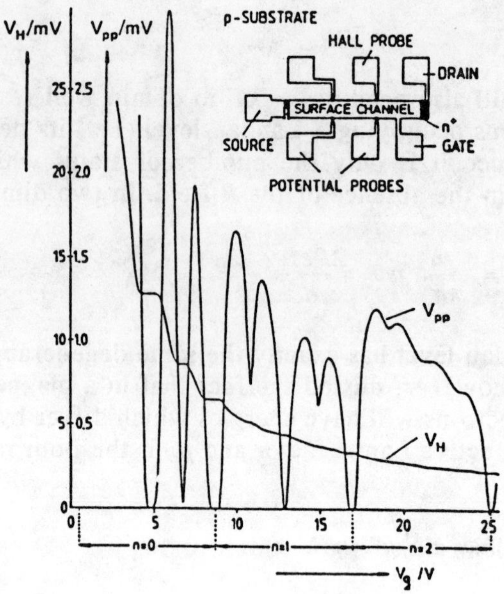

Fig. 10. Measured results for the Hall voltage, V_H, the voltage along the specimen illustrating the presence of zeros in the conductivity, σ_{xx}, and corresponding plateaus in the Hall resistance, p_{xy}. t is 1.5 K and B is 18 T from von Klitzing et al. (1980).

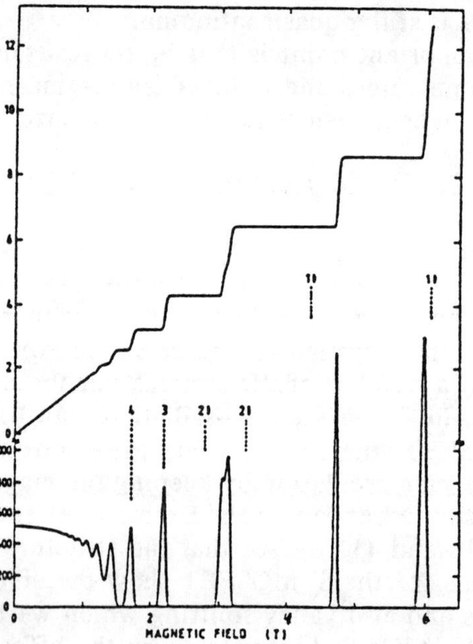

Fig. 11. Experimental curves for the Hall resistance, and for ρ_{xx} in a GaAs/AlGaAs heterostructure at about 8 mK. From von Klitzing, (1986).

is that conductivity seems to go to zero at some values of B and Vg. We might naively imagine that this corresponds to E_F lying between Landau levels so that $N(E_F)$ and thus σ is zero. The zeros correspond to E_F lying between Landau levels but $N(E_F)$ is not zero. Indeed, if the Landau levels were actually separated by gaps in the density of states all that would happen is that B or the electron density was varied E_F would jump from one Landau level to the next. When one level is full the addition of just one more electron moves E_F to the bottom of the next level.

To understand these zeros in conductivity we must briefly consider localization effects. Anderson first showed that if a conductor was sufficiently disordered all the previously delocalized states, which were extended throughout the material and carry currents, became localized within a region of the specimen. Mott extended this idea to show that if the disorder was less strong, states in the center of a conduction band would be delocalized (Figure 12). The delocalized and localized states are then separated in energy by a mobility edge, E_c so that for $E_F > E_c$ the conductivity would be metallic, i.e., σ is finite for $T = 0$, while for $E_F < E_c$ conduction can only occur by thermal activation.[1]

The electron states in the wings of the Landau levels are localized in the same way. As we see from Figure 12 the occupied delocalized states are then separated by a "mobility gap." This is similar to the situation in amorphous semiconductors where, due to the disorder, states exist in the normally forbidden energy gap, but these states are localized. As the Fermi level passes between Landau levels it does not jump from one to the next but for a finite range of electron densities it is in the mobility gap. Conductivity then tends exponential to zero as the temperature is lowered. If the energy difference between E_F and E_c is much larger than kT the probability of thermal activation, and thus conductivity, becomes vanishingly small.

8. The Quantum Hall Effect

In 1980 von Klitzing, Dorda, and Pepper showed that, in conditions of high magnetic field and low temperature, the Hall resistance, ρ_{xy} of a number of Si MOSFETs had unexpected plateaus over the ranges of carrier concentration at which the conductivity in the direction of the current, σ_{xx}, was close to zero (Figure 10). The values of ρ_{xy} on these plateaus were found to be independent of the specimen used and to be given by h/ie^2 for E_F lying in the mobility gap just above the ith Landau level. In this original

[1] Strictly speaking a mobility edge does not exist in 2D. However E_c can then be usefully considered as separating very weakly and very strongly localized states.

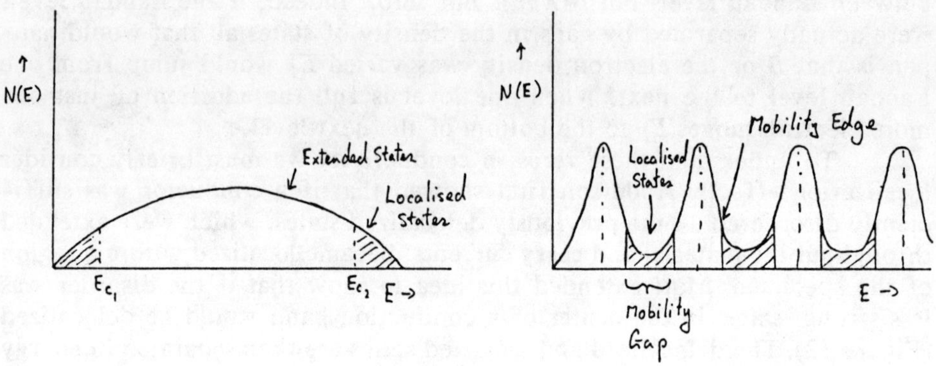

Fig. 12. Illustration of localization of states at the edges of bands and Landau levels.

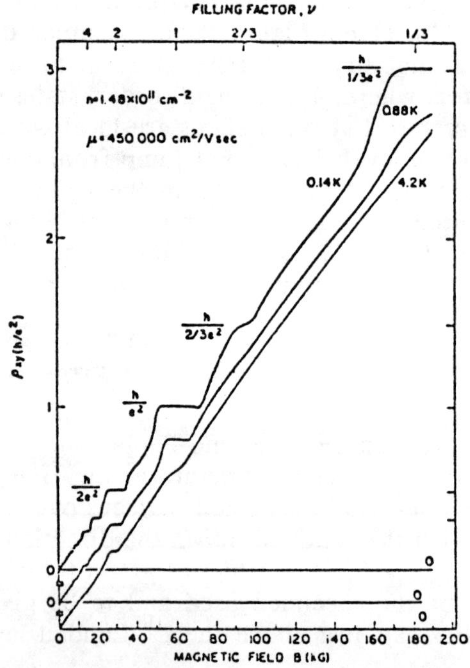

Fig. 13. Magnetic field dependence of the Hall resistance of a high mobility GaAs/AlGaAs heterostructure showing fractional quantization. From Tsui et al. (1982).

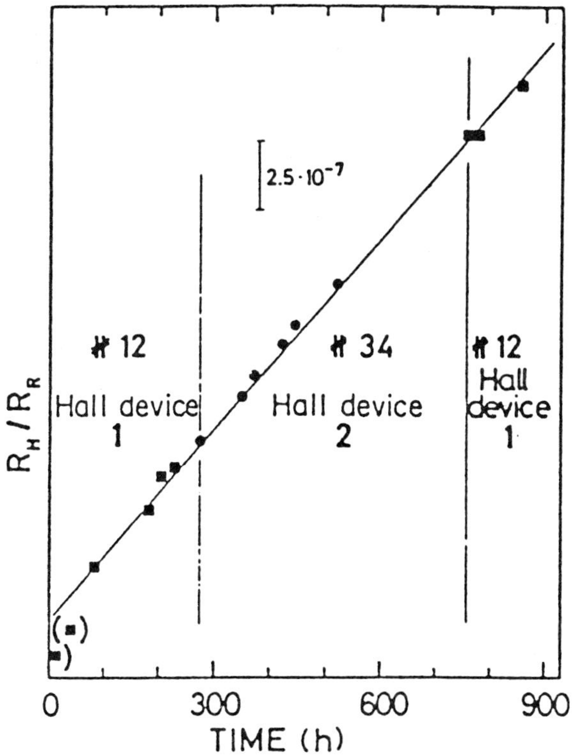

Fig. 14. Ratio of the quantum Hall resistance, R_H, and a standard wire resistor, R_R, as a function of time. From von Klitzing (1986).

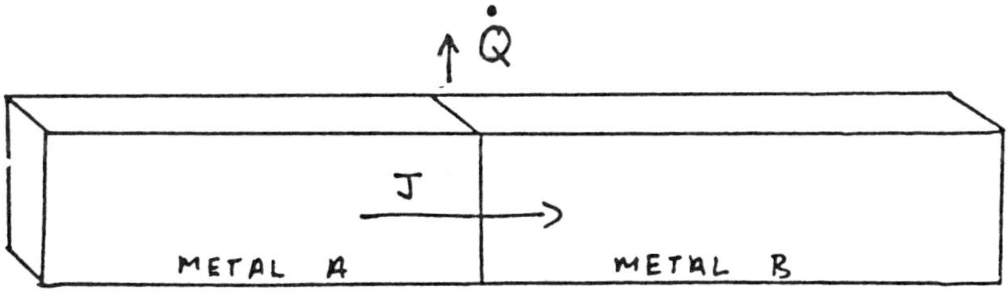

Fig. 15. Illustration of the Peltier effect. The current J passing through the junction of metals A and B produces heat Q.

paper they found that the values obtained agreed with the accepted value of $h/^{e2}$ to three parts per million. Subsequent work has shown that the agreement is within a few parts in 10^8! In Si MOSFETs we must use very high fields to observe these plateaus. In GaAs/AlGaAs heterostructures, because of the smaller effective mass and much higher mobilities, we can see well defined plateaus at relatively low magnetic fields. Figure 11 shows a series of very clear plateaus for such a heterostructure at a very low temperature. In this case the electron density is constant and the magnetic field is varied. To obtain such accurately quantized behavior in a simple solid state device is remarkable.

To understand how this arises it is useful to first look at the behavior of free electrons in crossed electric and magnetic fields. For B in the z-direction and E in the x-direction the classical solutions are:

$$x = v\cos\omega_c t, \quad y = \frac{E}{B}v\sin\omega_c t. \tag{13}$$

Thus the effect of the electric field is to produce a steady motion with velocity E/B of the cyclotron orbit in the y-direction, i.e., normal to both E and B.

The quantum mechanical solution is given in Appendix 1. This shows that the eigenstates correspond to harmonic oscillator states which have a constant group velocity along the y-axis equal to the classical value of E/B. For E_F just above the ith filled Landau level we will therefore have a current along the y-axis.

$$J_y = -nev_y = i\left(\frac{e^2B}{h}\right)\left(\frac{Ex}{B}\right) = i\,\frac{e^2}{h}\,Ex. \tag{14}$$

The Hall resistance defined as Ex/J_y is then h/ie^2, exactly the value obtained in the quantum Hall effect.

Thus, we can obtain the correct results for free electrons for filled Landau levels. This then leaves two major problems; first, our proof requires that there is no scattering of the electrons; second, in the experiment the Landau levels are not filled. The first problem is simply that scattering will allow transitions between harmonic oscillator states and thus a current in the x-direction, J_x. Classically the orbiting electrons can skip between orbits. if J_x is finite we have energy dissipation since $J{\cdot}E \neq 0$. Without scattering the current J_y will flow without dissipation.

If E_F lies in a mobility gap then at low enough temperatures the states which are mobile cannot be scattered since there are no available empty states to scatter into. Thus though $\sigma_{xx} \to 0$ as $T \to 0$, there is actually no electron scattering. We can therefore anticipate the result that the quantization is only exact in the limit $\sigma_{xx} \to 0$. Thus only the fraction of electrons within a Landau level which lie between the mobility edges contribute to the Hall coefficient. The plateaus exist therefore because as E_F moves through the mobility gap, the extra electrons drawn into the 2DEG populate localized states, and the density of mobile electrons stays fixed.

The reason the result seems to correspond to filled Landau levels is a good deal more subtle. It is closely connected to the fact that the Hall coefficient does not depend upon the geometry of the specimen in 2D in the absence of scattering. We did not need to make any reference to geometry in our derivation of the Hall coefficient. Thus if we take a specimen and physically punch holes through it we will still get the same result for R_H. All that happens is that the current density increases in the remaining regions. The effect of localized states can be shown in the same way; they are localized in certain regions of the specimen and these nonconducting areas act like holes in the specimen. The fraction of the mobile electrons occupies a corresponding fraction of the area, so the density of mobile states per unit area is exactly equal to the density of the filled Landau levels. We can thus construct a picture of the microscopic behavior which gives a plausible explanation of the QHE. The exact quantization suggests a more general explanation. This was provided by Laughlin (1981). The argument shows that quantization of the Hall coefficient h/ie^2 follows from flux quantization h/e and charge quantization e.

In 1982 Tsui, Stormer and Gossard investigated a very high mobility heterostructure at high fields and found plateaus corresponding to $i = 1/3$ and 2/3 (see Figure 13). Many further fractional values have now been observed, all of which have odd denominators. This fractional quantum Hall effect (FQHE) is only observed for very high mobilities. It is believed to be the result of electron-electron interaction, although it is far from well understood.

9. Application of the Quantum Hall Effect in Metrology

The quantity h/e^2 has the dimensions of electrical impedance and can be thought of as a quantum unit of resistance. Therefore, the QHE can be used as a resistance standard in the same way that the Josephson effect can be used to define a voltage standard in units of h/e. Furthermore, as pointed out in the original QHE paper by von Klitzing, Dorda and Pepper (1980), the fine structure content a is given by

$$\alpha = \left(\frac{\mu_0 c}{2}\right)\left(\frac{e^2}{h}\right) \tag{15}$$

and both μ_0 - the permeability of free space, and c - the speed of light, are defined units. Measurements of the QHE can therefore provide an independent method of measuring α. There are presently two ways to measure α, one of which depends on the correctness of quantum electrodynamics, QED. In principle this could therefore provide a test of QED. This of course assumes that the quantum Hall resistance R_{QH} is exactly given by h/e^2.

Experimentally, the accuracy with which R_{QH} can be measured is the same as that of any resistance. The best measurements to date seem to be those performed at the British National Physical Laboratory (NPL) and the Australian National Measurement Laboratory (NML) (see Sloggett et al. 1986). These are accurate to about 0.007 ppm, and are:

$$R_{\text{QH}} = 25812.8106\ (17)\ (\text{NPL})$$
$$R_{\text{QH}} = 25812.8099\ (20)\ (\text{NML})$$

while the value calculated from the accepted value of α is

$$R_\alpha = 25812\text{-}8036\ (22)$$

R_{QH} and R_α are therefore consistent with the 2-standard-deviation level. It is believed that the accuracy with which R_{QH} can be measured will be improved to the 0.01 ppm level in the next few years. If a real discrepancy is then apparent it may not be clear which method is giving the incorrect result. Quantization has only been proved to be correct for infinite specimens at zero temperature. I am currently investigating these possible corrections in collaboration with NPL.

Regardless of the exact quantization, the QHE can still be used to calibrate standard resistors since R_{QH} is far more stable and reproducible than any wire resistor (Figure 14). In this experiment the ratio of R_{QH} to the resistance of a standard reference resistor R_R is measured. This shows that R_R is slowly changing over time and that the value of R_{QH} shows no significant sample dependence. So R_{QH} may become accepted as the absolute resistance standard in the near future.

10. The Thermopower of the 2DEG

Following the discovery of the quantum Hall effect it was shown theoretically that the thermopower of the 2DEG in a quantizing magnetic field should show oscillatory behavior of a universal nature (see Johnson and Girvin, 1984). While the Hall resistance is only dependent upon e^2/h, thermopower was predicted to have peaks which in certain limits depend only upon k/e, where k is Boltzmann's constant. This has made the almost totally neglected area of the thermopower of the 2DEG a very active area of study.

If we apply a temperature gradient along the x-axis of a specimen, then an associated electrical field will be produced. The Seeback coefficient, which is often loosely referred to as "the thermopower" is then defined by $S_{xx} = E_x/\nabla T_x$ and the Nernst-Ettingshausen coefficient by $S_{xy} = E_y/\nabla T_x$. To get a feel for what thermopower measures it is best to consider the Peltier coefficient, π. From thermodynamics these coefficients are related by $\pi \equiv ST$. Consider the situation shown in Figure 15. The difference in the Peltier coefficients of the conductors A and B, π_{AB}, is defined as the heat generated or absorbed at their junction per unit area per unit current, i.e., $Q = \pi_{AB}J$. The Peltier coefficient of a conductor is thus the heat energy per unit charge brought to the junction and S is therefore the entropy per carrier. For nondegenerate carriers we therefore expect $\pi \cong kT/e$ and $S \cong k/e$. In the degenerate case only a fraction of carriers within kT of E_F carry heat energy. We then have $S \cong (kT/E_F)\cdot k/e$; which reflects the lower entropy. For activated conduction the heat energy is simply the activation energy, ΔE, than $\pi \cong \Delta E/e$ and $S \cong \Delta E/Te$. In making these estimates we have ignored scattering. If the hotter electrons are scattered more than the cooler electrons the rate at which heat energy is brought to the junction will be less. The thermopower therefore also depends upon the energy derivative of the relaxation time. In 2D the full form of the thermopower is

$$S = -\frac{\pi^2}{3}\left(\frac{kT}{E_F}\right)\frac{k}{e}(p+1) \text{ for metallic conduction} \tag{16}$$

and

$$S = -\frac{k}{e}\left(\frac{\Delta E}{kT}\right) + (p+2) \text{ for activated conduction} \tag{17}$$

where $\tau(E) \propto EP$. Thermopower is thus a good way of distinguishing between conduction mechanism.

The calculation of S in the quantum Hall regime is rather complicated. however, the general features are quite simply obtained. First, if E_F lies in the mobility gap then the carriers have no thermal energy and in the limit $\sigma_{xx} \to 0$ both S_{xx} and S_{xy} tend exponentially to zero. This is similar to the case of a superconductor where S is also zero. We can obtain the result for E_F in the center of the ith Landau level in the limit that $\hbar\omega_c \gg kT \gg \Gamma$. In this case the Landau levels are well separated and therefore only the carriers in the ith level have any thermal energy or entropy. The probability of occupancy of a state within this level when E_F is at the center will be $\frac{1}{2}$, since $kT \gg \Gamma$, the entropy per electron within the level will then be $kln2$. The entropy divided by the total number of electrons will then have the maximum value of $kln2\ (1 + \frac{1}{2})^{-1}$ and the maximum thermopower is therefore given by

$$S_{xx,\max} = \frac{k}{e}\frac{ln2}{(i + \frac{1}{2})}. \tag{18}$$

We thus expect the behavior shown in Figure 16. In the absence of scattering S_{xy} is zero, but if scattering is present S_{xy} is nonzero and it will also show oscillatory behavior.

This general behavior seemed to be confirmed by early experimental studies (see Obloh et al. (1984) for example), but more recent work led to very different results (see Fletcher et al. (1986) for example). The reason for the observed nonuniversal behavior is almost certainly due to phonon-drag.

In a temperature gradient there is net flux of phonons and hence a net phonon momentum. Electron-phonon scattering then leads to a transfer of momentum into the electron system and an electrical field along the specimen must arise to stop the current flow, resulting in a phonon-drag thermopower S_d. We might expect the phonon-drag thermopower in the case of the 2DEG to be very small because of the problem of conserving momentum in a collision between a 3D phonon and a 2D electron. This is not the case because for confinement in a well of width $\sim\delta$ the electron has a larger uncertainty in its momentum of $\sim 2\pi/\delta$. The phonon drag can in fact be strongly enhanced (Figure 17). If the phonons only lose momentum to the electrons they exert a pressure, $p = U(T)/3$, where $U(T)$ is the phonon energy density. So in a temperature gradient there is a net force given by

$$F_x = -\frac{dp}{dx} = -\frac{1}{3}\frac{dU}{dT}\cdot\frac{dT}{dx} = -\frac{Cv}{3}\frac{dT}{dx} \tag{19}$$

where Cv is the lattice specific heat. For n_A electrons per unit area the number per unit volume is m/t, where t is the thickness of the specimen, and at equilibrium we must have

$$\frac{n_A}{t}eE_x = -F_x = \frac{Cv}{3}\frac{dT}{dx} \tag{20}$$

$$\therefore S_d = \frac{E_x}{(dT/dx)} = \frac{tCv}{3n_Ae}. \tag{21}$$

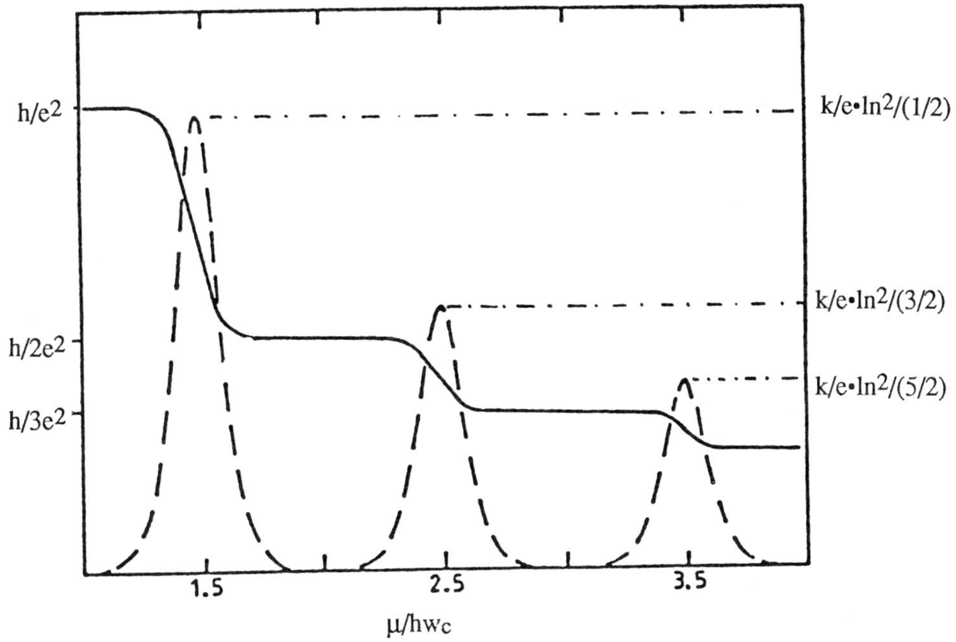

Figure 16. Calculated thermopwer, S_{xx} (broken curve) and Hall resistance, R_{xy}, (full curve) as a function of the reduced chemical potential, $\mu/h\omega_c$. From Streda (1983).

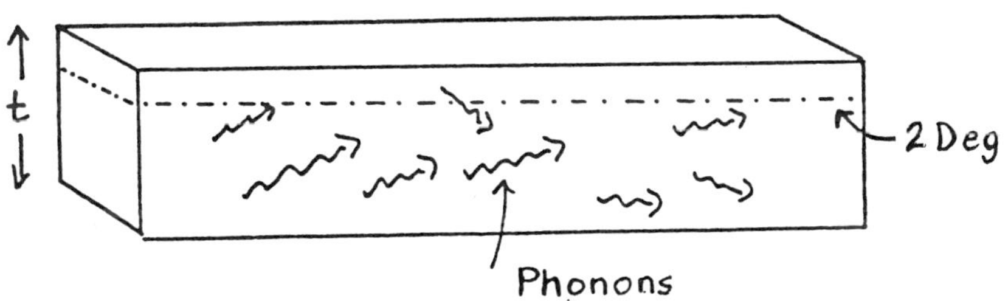

Figure 17. Illustration of the origin of photon drag.

The assumption that the phonons give all their momentum to the electrons is, of course, quite unrealistic. At low temperatures we can approximate the fraction of phonon momentum given to the electrons by the probability of a phonon being absorbed by the 2DEG in one pass through the 2DEG, f. The assumption here is that if it is not absorbed by the 2DEG it will lose its momentum on collision with the boundaries of the specimen. We then have

$$S_d \cong \frac{f\, t\, Cv}{3n_A e}. \tag{22}$$

As will be shown later f has very little temperature dependence at low temperatures and so we expect a T^3 dependence rather than the T dependence of the normal diffusion thermopower. Using Equations (16) and (21), it is simple to show that we expect $S_d/S \cong 10^3 \cdot Cv/C_el$, where C_el is the electronic specific heat. We therefore expect S_d to dominate except at very low temperatures. It is surprising therefore that the earlier work in this field ignored this contribution.

Calculation of S_d in a magnetic field has not, to my knowledge, been attempted. However, as pointed out by Nicholas (1985), it should show a structure which is qualitatively similar to that of the diffusion thermopower. When E_F lies in a mobility gap we expect zero contribution since the absence of available states to scatter into means that momentum cannot be absorbed. Similarly, when E_F is at the center of a Landau level the number of available final states in a maximum and S_d is a maximum.

In a recent paper Gallagher et al. (1987) showed that phonon-drag completely dominates the thermopower of Si inversion layers and that should also be the case in other 2D systems. At high electron densities S is roughly proportional to T^3 as can be expected (Figure 18). At low electron densities E_F passes through a mobility edge and then conduction becomes activated but phonon drag still seems to dominate. At the lowest electron densities conduction could be diminished by variable range hopping. This gives a $T^{1/3}$ dependence which is consistent with our results. The behavior at high electron densities is not exactly T^3 as shown in Figure 19. The peaks in these curves occur when the dominant phonon wavevector q_D, i.e., the position of the peak of the phonon distribution, is equal to $2k_F$. As q_D approaches $2k_F$ electron-phonon scattering increases due to the increase in the density of final states for the electrons to scatter into. As q_D becomes larger than $2k_F$ the electrons can only couple with the fraction of phonons with $q \leqslant 2k_F$. Thus S increases more rapidly than T^3 as $q_D \rightarrow 2k_F$ and then at high temperatures the power of T will rapidly fall. As can be seen, agreement with the full theory is rather good.

This cut-off in electron-phonon coupling can be seen very clearly in the results of hensel et al. (1983). They directed heat pulses at the 2DEG in

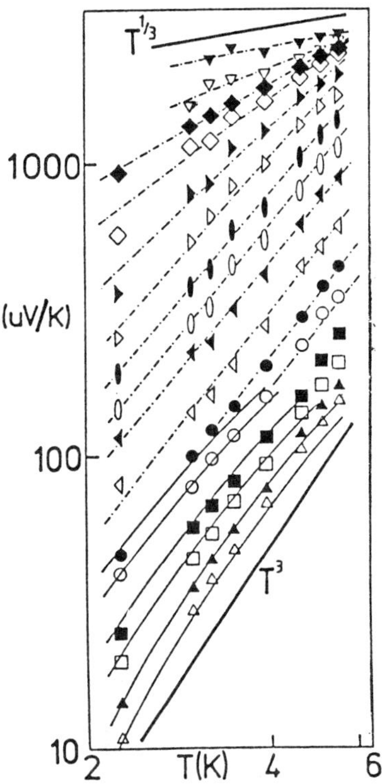

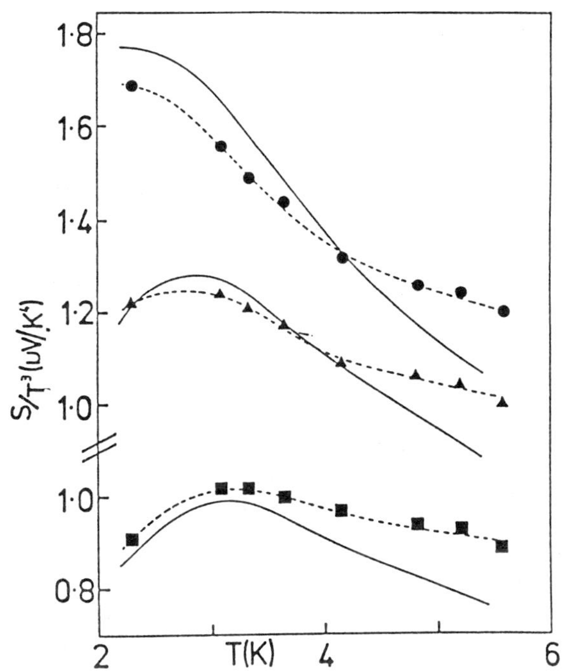

Fig. 18. The measured thermopowers of an Si inversion layer and the calculated phonon-drag thermopower (solid lines). From Gallagher et al. (1987). The different symbols correspond to different electron densities. S is smallest for the largest electron densities

Fig. 19. The reduced thermopower S/T^3 as a function of T. The solid lines are the calculated values. The electron densities are ■10^6, △ $8{\cdot}10^{15}$, and • $6{\cdot}10^{15}$ m^{-2}.

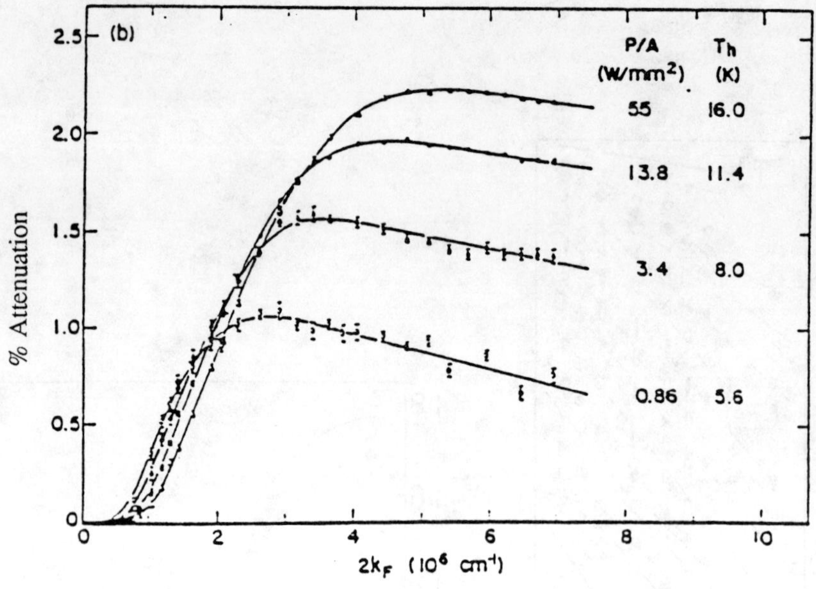

Fig. 20. The attenuation of a phonon pulse due to passage through the 2DEG of an Si MOSFET. From Hensel et al. (1983).

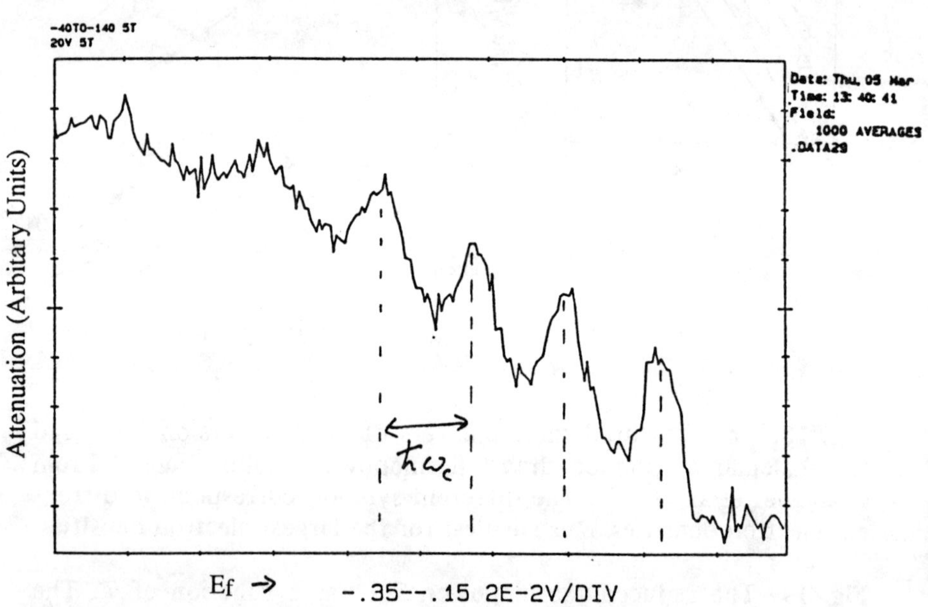

Fig. 21. The oscillatory structure seen in the attenuation of a phonon pulse in a *B* field. From Newton et al. (1987).

an Si MOSFET and looked for the fractional absorption as a function of gate voltage and hence electron density. As can be seen from Figure 20, the absorption at first increases in proportion to the electron density but then falls off rapidly for $2k_F > q_D$. Measurements of this kind are clearly closely related to the phonon drag thermopower. Very recent work on the absorption when a normal magnetic field is present has shown the presence of the expected oscillator behavior as E_F is swept through the Landau levels (Figure 21).

Thermopower measurements therefore have a great potential for elucidating the details of conduction in the 2DEG. If the diffusion and phonon-drag contributions can be successfully separated out, then it is possible to gain a good deal of information on the details of the quantum Hall effect and also on the electron-phonon interaction in the 2DEG.

11. Conclusion

In this brief review I have only been able to touch on a few aspects of electrical transport in 2D. I have said very little about superlattices, or electrical transport in quasi-1D systems. A listing of review articles covering many other aspects of low-dimensional semiconductor structures is given in Appendix B. The study of low-dimensional structures continues to produce a range of strikingly new phenomena. This has been one of the most interesting and productive areas within solid state physics in the last few years. I feel we can be confident that this will continue to be the case in the years ahead.

Appendix A

Solution of Schrödinger's Equation for Crossed Electric and Magnetic Fields

For a 2D free electron gas in the x-y plane with a B field in the z-direction and an E field in the x-direction the hamiltonian is that of Equation (8) plus the additional electrostatic potential energy, i.e.,

$$H = \frac{(p + eA)^2}{2m} - eEx. \tag{A1}$$

Rearranging we obtain

$$H = \frac{p_x^2}{2m} + \frac{1}{2} m\omega_c^2(x + x_0)^2 - eEx_0 + \frac{1}{2}m\left(\frac{E}{B}\right)^2 \tag{A2}$$

where $x_0 = py/m\omega_c + (E/B)$.
The eignevalues of an electron with this Hamiltonian are

$$E = (n + \tfrac{1}{2})\hbar\omega_c - eEx_0 + \tfrac{1}{2}m(\tfrac{E}{B})^2. \tag{A3}$$

This is the allowed energy states of a harmonic oscillator centered on x_0 ($-eEx_0$ being its electrostatic energy) moving along the y-axis with a group velocity $v_y = E/B$ ($\frac{1}{2} mv_y{}^2$ being its kinetic energy). Since the group velocity is

$$\begin{aligned} v_y &= \frac{1}{h}\frac{\partial\varepsilon}{\partial ky} = \frac{1}{h}\frac{\partial\varepsilon}{\partial x_0}\cdot\frac{\partial x_0}{\partial ky} = -\frac{1}{h}eE\left(\frac{h}{m\omega_c}\right) = -\frac{E}{B} \\ v_x &= \frac{1}{h}\frac{\partial\varepsilon}{\partial kx} = 0. \end{aligned} \tag{A4}$$

Appendix B

Relevant Reviews on Low-Dimensional Semiconductor Structures

Device Fabrication

Wood, C. E. C. (1980), Progress, Problems and Application of Molecular Beam Epitaxy, *Physics of Thin Films*, 11, 35-103.

Change, L. L. (1980), Molecular Beam Epitaxy, In: *Handbook of Semiconductors*, vol. 3, ed. by Keller (North Holland, Amsterdam), 563 pp.

Parker, E. H. C. (1985), ed., *The Technology and Physics of Molecular Beam Epitaxy*, (Plenum Press, NY and London).

Joyce, B. A. (1985), Molecular Beam Epitaxy, *Rep. Prog. Phys.*, vol. 48, pp. 1637-1699.

Mullin, J. B., et al. (1984), eds., Metallorganic Vapor Phase Epitaxy, *J. Cryst. Growth*, vol. 68.

Ploog, K., Dohler, G. (1983), Compositional and Doping Superlattices in III-V Semiconductors, *Adv. Phys.*, vol. 33, pp. 285-360

Kelly, M. J., Weisbuch, C. (1986), eds., The Physics and Fabrication of Microstructures and Microdevices, Springer, *Proc. in Physics*, vol. 13.

Physics of LDS

Ando, T., Fowler, A. B., Stern, F. (1982), Electronics Properties of Two-Dimensional Systems, *Rev. Mod. Phys.*, vol. 54, pp. 437-625.

Pepper, M. (1985), Localization and Quantization in Silicon Inversion Layers, *Contemp. Phys.*, vol. 26, pp. 257-293.

Bauer, G., Kuchar, F., Heinrich, H. (1984), eds., Two-Dimensional Systems, Heterostructures and Superlattices, Springer, *Solid State Science*, vol 53.

Kelly, M. J., Nicholas, R. J. (1985), The Physics of Quantum Well Structures, *Rep. Prog. Phys*, no. 48, pp. 1699-1741.

von Klitzing, K. (1986), The Quantum Hall Effect, *Rev. Mod. Phys.*, vol. 58, pp. 513-533.

Weisbuch, C. (1986), Fundamental Properties of III-V Semiconductor 2-D Quantized Structures: The Basis for Optical and Electronic Device Applications, In: *Semiconductors and Semimetals*, ed. by R. Williams and A. C. Beer (Academic Press).

References

1. Ando, T., Fowler, A. B., Stern, F., *Rev. Mod. Phys.*, vol. 54, no. 2, p. 437, 1982.
2. Broers, A. N., In: The Physics and Fabrication of Microstructures and Microdevices, Springer: *Proc. in Physics*, vol. 13, ed. by M. J. Kelly and C. Weisbuuch, p. 2. 1986.
3. Capasso, F., *Surf. Sci.*, vol. 142, p. 513, 1983.
4. Chang, L. L., Molecular Beam Epitaxy, In: *Handbook of Semiconductors*, ed. by S. P. Keller, Amsterdam: North Holland, p. 563, 1980.

5. Fletcher, R., Mann, J. C., Ploog, K., Weimann, G., *Phys. Rev. B*, vol. 33, p. 7122, 1986.
6. Gallagher, B. L., Gibbings, C. J., Pepper, M., Cantrell, D. G., *Semicon. Phys. and Tech.*, (in press), 1987.
7. Hensel, J. C., Dynes, R. C., Tsui, D. C., *Phys. Rev. B*, vol. 28, p. 1124, 1983.
8. Jonson, M., Girvin, S. M., *Phys. Rev. B*, vol. 29, p. 1939, 1984.
9. Laughlin, R. B., *Phys Rev. B*, vol. 23, p. 5632, 1981. See also, Laughlin, R. B. In: Two-Dimensional Systems, Heterostructures, and Superlattices, *Springer Series in Solid State Science*, No. 53, ed. by G. Bauer, F. Kuchar, and H. Heinrich, p. 272, 1984.
10. Mullin, J. B., Irvine, S. J. C., Moss, R. H., Robson, P. N., Wright, D. R., *J. Cryst. Growth*, vol. 68, 1984.
11. Narayanamurti, V, *Phys. Today*, vol. 37, p. 24, 1984.
12. Nicholas, R. J., *J. Phys. C*, vol. 18, p. L695, 1985.
13. Obloh, H., von Klitzing, K., Ploog, K., *Surf. Sci.*, vol. 142, p. 236, 1984.
14. Pepper, M., *Contemp. Phys.*, vol. 26, no. 3, p. 257, 1985.
15. Shah, N. J., Pei, S. S., Tu, C. W., Tiberio, R. C., *IEEE Trans. on Elec. Devices*, vol. 57, p. 3237, 1986.
16. Sloggett, G. J., Clothier, W. K., Ricketts, B. W., *Phys. Rev. Lett.*, vol. 57, p. 3237, 1986.
17. Smith, T. P., Stiles, P. J., *Solid State Commun.*, vol. 58, p. 511, 1986.
18. Stern, F., Howard, W. E., *Phys. Rev.*, vol. 163, p. 816, 1967.
19. Stormer, H. L., *Surf. Sci.*, vol. 132, 519, 1983.
20. Tsui, D. C., Stormer, H. L., Gossard, A. C., *Phys. Rev. Lett.*, vol. 48, p. 1559, 1982.
21. von Klitzing, K., Dorda, G., Pepper, M., *Phys. Rev. Lett.*, vol. 45, p. 494, 1980.
22. von Klitzing, K., *Rev. Mod. Phys.*, vol. 58, p. 519, 1986.

THEORY OF HEAVILY DOPED SEMICONDUCTORS

Virul Sa-yakanit
Semiconductor Physics Research Laboratory
Physics Department, Faculty of Science
Chulalongkorn University, Bangkok 10500
Thailand

Abstract: **This paper reviews the density of states available to the electron in heavily doped semiconductors. The discussion focusses on the density of states in the tail of the band by using three theoretical approaches: the semiclassical theory of Kane, the quantum mechanical theory of Halperin and Lax, and the path integral approach developed by Sa-yakanit. A model for interpolation of the density of states in the main part of the band and the tail of the band is proposed.**

I. Introduction

The experimental and theoretical investigations on heavily doped semiconductors have marked considerable progress in the recent years.[1] This is due to the fact that such materials are technologically important. These semiconductors are used in tunnel diode, laser, thermoelectric devices, semiconductor catalysts, and some active components of solid-state systems.[2] A second reason for the interest is the significant progress towards understanding the disordered materials. One extreme case of heavily doped semiconductors is that of disordered alloys[3].

A serious study of heavily doped semiconductors started about 20 years ago.[4] It was stimulated by the invention of a number of devices in which an important role is played by electronic states that are produced in the forbidden band under the influence of impurities. The new property of these impure crystals is discovered when the techniques for preparing pure semiconductor single crystal are developed. While the gradual reduction of impurity concentration is achieved it has become clear that the technology of semiconductor devices requires not simply crystals of maximum purity but crystals with impurities introduced deliberately in precisely known amounts, which are called "doped semiconductors." [1]

In the low-doping system[1], as shown in Fig 1a. , the isolated impurity ion has been described using the hydrogenic model resulting in the introduction of a discrete energy level in the forbidden energy gap. As the number of impurity ions increase, eventually the hydrogen-like wavefunctions on adjacent ions overlap and interact causing the broadening of the discrete level (Fig. 1 b). For a doping concentration at which the discrete level is beginning to broaden the center of the impurity band is expected to be the same as the position of the discrete level. As one increases the doping, the screening length is increased due to the carriers from the ionized dopants. The whole band then shifts towards the adjacent main band in the same way that the discrete level moves toward the adjacent band. When the center of the impurity band merges into the main band, the two entities, become one with a parabolic main band plus a tail. This defined the "heavily doped", or, in other word, a semiconductor is considered heavily doped when the impurity band associated with the doped impurity merges with either the conduction or the valence band see Fig. 1 c.

Evidence for band tails in the forbidden band of heavily doped semiconductors was provided by the number of experiments on tunneling, optical absorption and luminescence. The electronic properties of heavily doped semiconductor depends largely on the tail in density of states. For this reason theories of impured semiconductors focussed on determining density of states in this tail using different theoretical approaches. However, the perturbation techniques lead to the tail that cut off sharply.

Kane[5] has combined the potential energy fluctuations with the Thomas-Fermi method or semi-classical method, to calculate the density of states. In this method he assumes that the local potential is sufficiently slowly varying that the local density of states can be defined just as if the potential were constant. The calculation of the over all density of states then reduces to the calculation of the distribution function. It is noticed that the potential energy fluctuations at high concentrations are Gaussian, the tail found by Kane is Gaussian. Owing to the omission of the kinetic energy of localization in the Thomas-Fermi method, the density of states obtained from Kane Theory lead to the overstimated value

$$\rho(E) = y(E/\eta)\, m^{*3/2} (2\eta)^{1/2}\, V\, \pi^{-2}\, \hbar^{-3} \tag{1.4}$$

where $y(x) = \pi^{-1/2} \int_{-\infty}^{x} (x-\xi)^{1/2} \exp(-\xi^2)\, d\xi$

The most complete calculation of density of states in the low energy, deep tail region of the impurity-band remains the optimal fluctuation results of Halperin and Lax theory.[6-7] Their theory is the quantum counterpart of the original semi-classical theory of Kane. Quantum effects were included by adding the zero point energy of the electron states (kientic energy of localization) which raises the electron state energies and reduced density of states at small energy below the semiclassical value obtained by Kane.

Recently Sa-yakanit[8-9] has used the Feynman path integral technique to acheive an analytical form of density of states. The density of states evaluated by this method agree with that evaluated by Halperin and Lax i.e.,

$$\rho(E) = [A(E)/\xi^2] \exp [-B(E)/2\xi] \tag{1.5}$$

The functions A(E) and B(E) evaluated by path integral method are expressed analytically in terms of parabolic cylinder function. Although, it is difficult to obtain the numerical values of this function, but it is more convenient to calculate the density of states by evaluating the parabolic-cylinder function than by numerical solving Hartree-Fock equation as was done by Halperin and Lax.

In this review paper the density of states available to the electrons in heavily doped semiconductor is discussed using the non-perturbative method such as the semiclassical theory of Kane, the quantum mechanical approach of Halperin and Lax in particularly the path integral method developed by Sa-yakanit . An application to the evaluation of the optical properties of heavily doped GaAs is given.

II Potential in Disorders System

An electron in a disordered solid sees a potential $V(\vec{r})$ that varies unexpectedly from point to point in the solid. The V is not zero as in the free electron case, it is not periodic as in the ordered solid case. The $V(\vec{r})$ could be due to the N ions which make up an amorphous solid so that

$$V(\vec{r}) = \sum_{i=1}^{N} v(\vec{r} - \vec{R}_i), \tag{2.1}$$

where $v(\vec{r} - \vec{R}_i)$ is the potential at $\vec{r}$ due to ion at point $\vec{R}_i$. $V(\vec{r})$ in (2.1) could also be due to N impurity ions substituted at random in an otherwise periodic solid. The $V(\vec{r})$ is then a fluctuating potential superimposed on a periodic potential.

In the present model the $\vec{R}_i$ are randomly distributed throughout the solid. That is, the probability of observing impurity ion i in volume element $d\vec{R}$ is just $d\vec{R}/\Omega$, where Ω is the volume. We can characterize V by its mean value V_o,

$$V_o \equiv \langle V(\vec{r}) \rangle = \int \frac{d\vec{R}_1}{\Omega} \cdots \frac{d\vec{R}_N}{\Omega} \sum_i v(\vec{r} - \vec{R}_i)$$

$$= N \int \frac{d\vec{R}}{\Omega} v(\vec{r} - \vec{R}) = n \int d\vec{R}\, v(\vec{r} - \vec{R}) \tag{2.2}$$

(all impurities identical) and its mean square value ξ .

$$\xi = n \int d\vec{R}\, v^2(\vec{r} - \vec{R}). \tag{2.3}$$

Here n $(= N/\Omega)$ is the impurity density.

The V_o and ξ set the mean and the magnitude of the fluctuation about the mean in the potential seen by the electron. We will also need the correlation of $V(\vec{r})$ at different points in the crystal which we denote by

$$W(\vec{r} - \vec{r}') \equiv \langle V(\vec{r})\, V(\vec{r}') \rangle - V_o^2$$

$$= n \int d\vec{R} \;\; v(\vec{r} - \vec{R})\, v(\vec{r}' - \vec{R}) \qquad (2.4)$$

Clearly from (2.3) and (2.4)

$$\xi = W(o) \qquad (2.5)$$

so that ξ measures the magnitude of the correlations.

We also consider the high density limit. This means we have many impurities lying within the wave function of a single electron. The electron wave function can extend several hundred angstroms and high concentration begins in practice at $n \sim 10^{17}$ centres/cm^3. We also consider weakly attractive impurities (v small and negative) such that nv^2 = constant. In this case $nv^3 \ll nv^2$, for example.

A specific representation of the impurity potential we will consider is the screened Coulomb potential.

$$v(\vec{r} - \vec{R}) = - \frac{e^2}{\varepsilon|\vec{r} - \vec{R}|} e^{-Q|\vec{r} - \vec{R}|}, \qquad (2.6)$$

where Q^{-1} is the screening length and ε the dielectric constant. In this case,

$$V_o = -n \frac{4\pi e^2}{Q^2}, \quad \xi = n \frac{2\pi e^4}{\varepsilon^2 Q}, \qquad (2.7)$$

and the correlation function is

$$W(\vec{r} - \vec{r}') = \xi \, e^{-Q|\vec{r} - \vec{r}'|} \tag{2.8}$$

However for the moment we leave v arbitrary with V_o, ξ and $W(\vec{r} - \vec{r}')$ parametrizing its properties.

III Kane Theory

Before considering the Kane theory let us first consider the free electron model.

In this case $V = 0$ and the electron wave functions are plane waves $\phi_{\vec{k}} = \Omega^{-\frac{1}{2}} e^{i\vec{k}.\vec{r}}$ corresponding to momentum $\hbar \vec{k}$ and energy $E_k = \hbar^2 k^2/2m^*$. The density of states per unit volume at energy E is defined as

$$\rho(E) = \frac{1}{\Omega} \sum_{\vec{k}} \delta(E - E_k) \tag{3.1}$$

where the sum is over the free electron energies E_k. Taking large we may convert the sum over $\vec{k}$ to an integral

$$\sum_{\vec{k}} \longrightarrow \frac{\Omega}{(2\pi)^3} \int d^3k$$

and

$$\rho(E) = \frac{1}{(2\pi)^3} \int_0^\infty 4\pi k^2 dk \delta(E - E_k)$$

$$= \frac{1}{(2\pi)^2} \left(\frac{2m^*}{\hbar}\right)^{3/2} \int_0^\infty dE_k \, E_k^{\frac{1}{2}} \, \delta(E - E_k) \tag{3.2}$$

To use the properties of the delta function

$$\int_{-\infty}^{\infty} dx \, f(x) \, \delta(x - a) = f(a)$$

we introduce the Heaviside step function

$$H(E) = \begin{cases} 1 & E > 0 \\ 0 & E < 0 \end{cases} \tag{3.3}$$

so that

$$\rho(E) = \left(\frac{m^{*3/2}}{2\pi^2\hbar^3}\right) \int_{-\infty}^{\infty} dE_k \, E_k^{\frac{1}{2}} \, H(E_k)\delta(E-E_k)$$

and

$$\rho(E) = C \, E^{\frac{1}{2}} \, H(E) \tag{3.4}$$

where

$$C = \left(\frac{m^{*3/2}}{2\pi^2\hbar^3}\right) \tag{3.5}$$

Hence $\rho(E) \propto E^{\frac{1}{2}}$ measured away from a potential $V = 0$.

Kane's theory is a semiclassical treatment of an electron in a disordered solid. This means we ignore the zero point energy of the electron (a quantum effect). As noted above, the potential $V(\vec{r})$ varies from point to point in the solid. In regions where $V(\vec{r})$ is deep, we expect the electron wave function to be narrow (a highly localized electron) and the kinetic energy will be large in these regions. Conversely when $V(r)$ is nearly zero and flat, the kinetic energy is small (and slowly varying with $\vec{r}$).

The central assumption of Kane's theory is to ignore the variation of kinetic energy with $V(\vec{r})$ and assume the electrons can be treated as "free electrons" having the same kinetic energy everywhere in the solid for all $V(\vec{r})$. The $V(\vec{r})$ then simply sets the potential level from which we measure the energy of these "free electrons". The density of states at point $\vec{r}$ is then proportional to $\sqrt{E - V(\vec{r})}$. Explicity, using the free electron result,

$$\rho(E - V) = C\sqrt{E - V}\; H(E - V) \tag{3.6}$$

This $\rho(E - V)$ can be regarded as a function of V or (through $V(\vec{r})$). The density of states for the solid is then (3.6) averaged over all points in the crystal (or equivalently all potential energy values V). That is,

$$\rho(E) = \langle \rho(E - V) \rangle_V \tag{3.7}$$

where

$$\langle \, \rangle_V = \int dV\, P(V) \tag{3.8}$$

and $P(V)$ is the probability distribution of V, giving the probability of observing V in the range of V to $V + dV$. With this central assumption it remains only to determine $P(V)$ to complete the calculation of $\rho(E)$. Ignoring the variation of the kinetic energy with $V(\vec{r})$ will be valid in practice when V is long range and slowly varying in space. For weak and dense scatters, we can shown that

$$P(V) = \frac{e^{-\frac{1}{2}(V-V_o)^2/\xi}}{\int_{-\infty}^{\infty} dV\, e^{-\frac{1}{2}(V-V_o)^2/\xi}} \qquad (3.13)$$

The expressions (3.11) and (3.13) show that $P(V)$ is a Gaussian distribution about the mean V_o. A potential having this distribution is called a "Gaussian Random Potential". This distribution follows (1) from assuming the impurities are equally likely to be in any volume element in the solid and (2) from retaining quadratic terms in V only.

Using (3.13) for $P(V)$ in (3.7) and using the Heaviside function to reduce the integration up to E) we have

$$\rho(E) = \int_{-\infty}^{\infty} dV\, P(V)\; C\sqrt{E - V}\, H(E - V)$$

$$= (2\pi\xi)^{-\frac{1}{2}}\; C \int_{-\infty}^{E} dV \sqrt{E - V}\; e^{-\frac{1}{2}(V - V_o)^2/\xi} \qquad (3.14)$$

This is Kane's result for the density of states.

We consider two limits of (3.14). For high energy $E >> V$ we can ignore V (i.e. set $V = 0$) and take $E \to \infty$. Then

$$\rho(E) = C(2\pi\xi)^{-\frac{1}{2}}\, E^{\frac{1}{2}} \int_{-\infty}^{\infty} dV\, e^{-\frac{1}{2}(V - V_o)^2/\xi}$$

$$= C\, E^{\frac{1}{2}} \qquad (3.15)$$

Thus Kane's result for high electron energies, where V is negligible goes over to the free electron value-as expected. For $E << V$, that is electron energies deep in the potential region, we have

$$\rho(E) = \frac{C}{2\sqrt{2}} \xi^{\frac{1}{4}} e^{-\frac{1}{4}(E-V_o)^2/\xi} D_{-3/2}\left(-\frac{(E-V_o)}{\sqrt{\xi}}\right) \tag{3.16}$$

where $D_{3/2}$ is the parabolic cylinder function .

Finally, since $E - V_o$ is large and negative, we use the asymptotic expression for $D_{-3/2}$ valid at large argument,

$$D_p(x) \approx e^{-x^2/4} x^p,$$

so that

$$\rho(E) = \frac{C}{2\sqrt{2}} \xi (V_o - E)^{-3/2} e^{-\frac{1}{2}(E-V_o)^2/\xi} \tag{3.17}$$

This is Kane's central result for the density of states at low energy in the "deep tail region". We see that $\rho(E) \propto e^{-E^2}$ has an exponentail tail at low energies. This tail falls of as E^2 which we shall see is valid in practice for long range potential v only. The existence of the tail in $\rho(E)$ is the new feature characteristic of a disordered system.

Finally, to make contact with Kane's original expressions we introduce the dimensionaless energy and density of states variables used by him,

$$x = \frac{(E - V_o)}{2\xi} \qquad y = \rho(E)/[2C(2\xi)^{\frac{1}{4}}]$$

The $\rho(E)$ in (3.16) is then

$$y(x) = \frac{1}{2^{11/4}} e^{-x^2/2} D_{-3/2}(-\sqrt{2}x) \tag{3.18}$$

which is known as "Kane's function". and can be compared with (1.4). The asymptotic expression for $D_{-3/2}$ is

$$D_{-3/2}(-\sqrt{2}x) \sim \begin{cases} e^{x^2/2} \quad x^{\frac{1}{2}} & x \gg 1 \\ e^{-x^2/2} \quad x^{-3/2} & x \ll -1 \end{cases} \tag{3.19}$$

lead to the asymptotic forms

$$y \sim \begin{cases} \sqrt{x} & x \gg 1 \\ x^{-3/2}\, e^{-x^2} & x \ll -1 \end{cases} \tag{3.20}$$

The dimensionless density of states y is sketched in Fig. 2.

IV The Halperin and Lax Theory

In Kane's semiclassical theory the kinetic of the electrons is assumed independent of the potential $V(\vec{r})$, In fact, in the regions of deep narrow potential wells the electrons will be highly localized and the kinetic energy of localization will be large. In regions of shallow or wide potentials the electron kinetic energy will be small. Hence the complete electron energy states, kinetic plus potential, will in general be higher than with the kinetic energy ignored, particularly in regions of deep wells. This is depicted schematically in Fig. 3.

Halperin and Lax set out to include the kinetic energy in a full quantum mechanical calculation of the density of states. In this case the expression (3.6) used by Kane must be abandoned and replaced by

$$\rho(E) = \frac{1}{\Omega} \sum_i \langle \delta[E - E_i(V)] \rangle_V \tag{4.1}$$

where $E_i(V)$ are the "exact" energy states of the electron in a potential V, including the kinetic energy. The average is again given by (3.8) and (3.13). In this case we use the central limit theorem to justify using the Gaussian potential distribution function P(V).

To evaluate (4.1) Halperin and Lax first emphasize that they are interested in $\rho(E)$ for low E in the band tail region. In this case only the lowest possible energy states found in regions of deep wells in $V(\vec{r})$ need be included in (4.1). Also they include only the ground states of the electrons in these deep wells. Secondly, at low E only a few states E_i will be low enough to satisfy the delta function in (4.1). As noted these will occur in regions of the crystal where the potential $V(\vec{r})$ has a deep well. However, if $V(\vec{r})$ is very deep it will also be narrow so that the kinetic energy will be large. Thus Halperin and Lax assume that for a given E there is an optimum well depth to get the most values of low E_i. They denote the wave function $\psi_i(\vec{r})$ for this optimum well by

$$\psi_i(\vec{r}) = f(\vec{r} - \vec{y}_i)$$

where $\vec{y}_i$ is the centre of $f(\vec{r} - \vec{y}_i)$. A one to one correspondence between the E_i and $\phi_i(r)$ is assumed and all other well shapes and $\phi_i(\vec{r})$ are ignored (assumed to occur much less often). The $f(\vec{r} - \vec{y}_i)$ is assumed real and normalized. Since the Hamiltonian of the disordered solid is given by

$$H = T + V(\vec{r}),$$

where

$$T = -\frac{\hbar^2}{2m^*} \nabla^2 + V_o$$

includes the mean value V_o of the random potential characterized by (2.1), (2.2) and (2.3), the energy corresponding to $f(\vec{r} - \vec{y}_i)$ is

$$E(\vec{y}_i) = \int f(\vec{r} - \vec{y}_i) [-\frac{\hbar^2}{2m^*} \nabla^2 f(\vec{r} - \vec{y}_i)]d\vec{r} + V_o$$

$$+ \int f^2(\vec{r} - \vec{y}_i)V(\vec{r})d\vec{r}$$

$$= T_s + V_o + V_s(\vec{y}_i) \equiv \theta + V_s(\vec{y}_i) \qquad (4.2)$$

Here only the averaged potential $V_s(\vec{y}_i)$ depends on $\vec{y}_i$.

With the assumption that all E_i are given by the optimum $E(\vec{y}_i)$ we have

$$E_i \approx E(\vec{y}_i) = \theta + V_s(\vec{y}_i) \qquad (4.3)$$

and

$$\rho(E) = \frac{1}{\Omega} \sum_i \langle \delta[E - (\theta + V_s(\vec{y}_i))] \rangle \qquad (4.4)$$

Note that this density of states is effectively the same as used by Kane except that the kinetic energy θ corresponding to the smoothed $V_s(\vec{y}_i)$ is included and must be calculated for each $V_s(\vec{y}_i)$. The form of the smoothed potential is shown schematically in Fig.6..

At this point we do not know the functions $f(\vec{y}_i)$. However since the $E(\vec{y}_i)$ are intended to represent ground states $\vec{E}_i$ we can use a variational principle to find $f(\vec{y}_i)$. That is $E(\vec{y}_i)$ will always lie above E_i, and the best fit $f(\vec{y}_i)$ will make $E(\vec{y}_i)$ closest to E_i. (Also since the $E(\vec{y}_i)$ always lie above the true E_i the $\rho(E)$ calculated using the $E(\vec{y}_i)$ will always be too small at low E. Hence we could maximize $\rho(E)$ to find $f(\vec{y}_i)$. In this variation $\vec{y}_i$ is treated as a variational parameter. Hence the optimum values of $\vec{y}_i$ occur when $E(\vec{y}_i) = 0$ or, since only $V_s(\vec{y}_i)$ in (4.2) depends on $\vec{y}_i$, then

$$\nabla V_s(\vec{y}_i) = 0$$

There are many $\vec{y}_i$ which satisfy this condition and to deal with this we insert a Dirac delta funciton in (4.4) so that

$$\rho(E) = \int \frac{d\vec{y}}{\Omega} \langle \sum_i \delta(\vec{y} - \vec{y}_i) \delta[E - (\theta + V_s(\vec{y}))] \rangle_V \qquad (4.5)$$

After averaging, $\rho(E)$ is not a function of $\vec{y}$, or

$$\rho(E) = \sum_i \langle \delta(\vec{y} - \vec{y}_i)\, \delta[E - (\theta + V_s(\vec{y}))] \rangle_V \tag{4.6}$$

If we transform from the variable $\vec{y}$ to $\nabla V_s(\vec{y})$ by writing

$$\sum_i \delta(\vec{y} - \vec{y}_i) = \delta\,(\nabla V_s(\vec{y}))\ \det\ \left| \nabla\nabla V_s(\vec{y}) \right|$$

we have

$$\rho(E) = \langle \delta[E - (\theta + V_s(\vec{y}))]\, \delta[\nabla V_s(\vec{y})] \left|\det \nabla\nabla V_s(\vec{y})\right| \rangle_V \tag{4.7}$$

This expression for $\rho(E)$ is expressed entirely in terms of $V_s(\vec{y})$. Halperin and Lax then replace V by $V_s(\vec{y})$ in the averging and by a sophisicated analysis of Gaussian averaging obtain

$$\rho(E) = \frac{\sigma_1\sigma_2\sigma_3(\theta - E)^3}{(2\pi)^2\xi^2\sigma_o^7} \exp\left[- \frac{(\theta - E)^2}{2\xi\sigma_o^2}\right] \tag{4.8}$$

Here

$$\sigma_o^2 = \int f^2(\vec{r})\, f^2(\vec{r})\, W(\vec{r} - \vec{r}')d\vec{r}\, d\vec{r}'$$

where $W(\vec{r} - \vec{r}')$ is the correlation function of the potental given by (2.4). Note that from (2.5) $nW(o) = \xi$. The σ_1, σ_2 and σ_3 are the diagonal elements of $\nabla\nabla\ V_s(y)$.

To find the optimum wave function f, Halperin and Lax maximize $\rho(E)$. Since $\rho(E)$ is dominated by the exponent they approximate this by minimizing the exponent,

$$\frac{\partial}{\partial f}\left[\frac{\theta - E}{\sigma_o}\right]^2 = 0 \tag{4.9}$$

This leads to a Hartree-Fock like equaiton for f

$$Tf(\vec{r}) - \mu f(\vec{r}) \int f^2(\vec{r}\,') \; W(\vec{r} - \vec{r}\,')d\vec{r}\,' = Ef(\vec{r}) \tag{4.10}$$

which must be solved numerically. Since E is given, the parameter μ corresponds to the eigenvalue in (4.10) and must be determined in the solution of (4.10). Once f obtained the σ_o, its derivatives σ_1, σ_2, σ_3 and the mean kinetic energy θ can be calculated numerically for substitution in (4.8).

Halperin and Lax used the screened Coulomb potential (2.6) in the explicit calculation of $\rho(E)$. The result, expressed in terms of the dimensionless energy $\nu = -(E - V_o)/E_Q$ where $E_Q = \hbar^2 Q^2/2m^*$ is the energy associated with the screened Coulomb potential, is

$$\rho(E) = \frac{Q^3}{E_Q(\xi')^2} a(\nu) e^{-\frac{b(\nu)}{2\xi'}} \tag{4.11}$$

Here

$$\xi' = \frac{\xi}{E_Q^2} = \frac{n\, 8\pi m^{*2} Z^2 e^4}{\hbar^4 Q^5 \varepsilon^2} \tag{4.12}$$

is the variance expressed in units of E_Q and $a(\nu)$ and $b(\nu)$ are "universal dimensionless functions" that Halperin and Lax calculated numerically.

V. The Path Integral Theory

Although the Halperin and Lax $\rho(E)$ agreed well with experiment in the low energy band tail region and is a substantial improvement over Kane's $\rho(E)$, it has two limitations. Firstly, it requires numerical solution. Secondly, it cannot be extended to higher energy E. For these reasons Sa-yakanit developed a new theory based on the Feynman formulation of quantum mechanics. The Feynman formulation is useful since, although a full quantum theory, it expresses the electron propagation as a sum of classical like paths. This is especially useful in disordered systems for then the density of states can be calculated in much the same way as in the Kane theory. Once the Feynman method is mastered this is much simpler than the Schrodinger picture of quantum mechanics for disordered systems. The Feynman path integral approach was first applied to disordered systems by Edwards[12, 13] and has subsequently been explored by others[14, 18]. However, Sa-yakanit[8-9] succeeded in evaluating the path integrals explicitly and obtained an expression for $\rho(E)$ valid at all energies E. In the low energy limit the $\rho(E)$ reduces to a relatively simple analytic expression. At high E it reduces to the usual free electron value (3.4).

The path integral formulation begins with the full expression (4.1) for $\rho(E)$,

$$\rho(E) \quad = \quad \frac{1}{\Omega} \sum_i < \delta(E - E_i(V)) > v. \qquad (5.1)$$

Next the retarded propagator $G(\vec{r}_2, \vec{r}_1, t)$ is introduced which describes the propagation of an electron from point $\vec{r}_1$, to $\vec{r}_2$. (This $G(\vec{r}_2, \vec{r}_1, t)$ is $(-i\hbar)$ times the usual retarded Green function). The $\sum_i \delta(E - E_i)$ can be expressed in terms of the diagonal element $(\vec{r}_1 = \vec{r}_2)$ of G giving

$$\rho(E) = \frac{1}{\pi\hbar} \text{Re} \int_0^{\infty} dt\, e^{iEt/\hbar} \langle G(\vec{r}_1, \vec{r}_1, t) \rangle v \tag{5.2}$$

The G is then written as a path integral over all the possible paths between $\vec{r}_1$ and $\vec{r}_2$,

$$G(\vec{r}_1, \vec{r}_2, t) = \int D(\vec{r}(\tau))\, e^{i/\hbar\, S_\ell} \tag{5.3}$$

where S_ℓ is the action

$$S_\ell = \int_0^t d\tau \left[\frac{m^*}{2} \dot{\vec{r}}^2(\tau) - V(\vec{r}(\tau))\right]$$

Here $V(\vec{r}(\tau))$ is the potential given by (2.1). The average $\langle\ \rangle v$ over G can be performed exactly as in the Kane theory, giving

$$\overline{G} \equiv \langle G(\vec{r}_2, \vec{r}_1, t) \rangle v = \int D(\vec{r}(\tau))\, e^{i/\hbar S} \tag{5.4}$$

where

$$S = \int_0^t d\tau \left[\frac{m^*}{2} \dot{\vec{r}}^2(\tau) - V_0 + \frac{i}{2\hbar} \int_0^t d\tau'\, W(\vec{r}(\tau) - \vec{r}(\tau'))\right] \tag{5.5}$$

Note the similarity between the exponent of P(V) in (3.11) and this averaged S. Essentially in S the kinetic energy is included and ε is replaced by $W(\vec{r}\ \ \vec{r}')$, the correlation function of v.

To proceed in calculating G, the full action S is modeled by a non-local harmonic "trial" action[8]

$$S_o = \int d\tau\ \frac{m^*}{2}\left[\dot{\vec{r}}^2(\tau) - \frac{\omega^2}{2t}\int_0^t d\tau'\,|\,\vec{r}(\tau) - \vec{r}\,(\tau')\,|^2\right] \qquad (5.6)$$

Essentially $W(\vec{r} - \vec{r}')$ is modeled by a single harmonic well. With S replaced by S_o, G in (5.4) can be evaluated exactly. In S_o, G in (5.4) can be evaluated exactly. In S_o, ω^2 measures the steepness of the harmonic well and ω^2 is treated as a variational parameter. With S replaced by S_o the resulting G constitutes the zero order approximation G_o to G. The full G can also be re-arranged as

$$\overline{G} = \int D(\vec{r}(\tau))e^{i/\hbar S} \int D(\vec{r}(\tau))e^{\frac{i}{\hbar}S_o} \frac{\int D(\vec{r}(\tau))e^{\frac{i}{\hbar}[S_o + (S-S_o)]}}{\int D(\vec{r}(\tau))e^{i/\hbar\ S_o}} \qquad (5.7)$$

or in short form

$$\overline{G} = G_o <e^{-i/\hbar(S-S_o)}>o \qquad (5.8)$$

where < >o denotes the average with respect to the trial action.

The $< \ >_o$ may be expanded in cumulants and keeping only the first order cumulant we obtain

$$G_1 = G_o \exp [\tfrac{i}{\hbar} < S-S_o >] \tag{5.9}$$

For a screened Coulomb $v(\vec{r})$ for which $W(\vec{r} - \vec{r}')$ is given by (2.8) the diagonal part of G_1 amy also be evaluated exactly[17] and gives the density of states, when substituted into (5.2), of

$$\rho_1(E) = \frac{1}{\pi\hbar} \mathrm{Re} \int_0^\infty dt \, [(\frac{m^*}{2\pi i\hbar t})^{\frac{1}{2}} (\frac{\omega t}{2\sin(\frac{1}{2}\omega t)})]^3$$

$$\times \exp [- \frac{i}{\hbar}(V_o-E) t - \frac{1}{2\hbar^2} \frac{Qt}{\sqrt{\pi}} \int_0^t dx \int_0^\infty dy \, y e^{-Q^2 y} g(x,y)^{-3/2}$$

$$+ \frac{3}{2} (\omega\frac{t}{2} \cot \frac{\omega t}{2} - 1)] \tag{5.10}$$

where

$$g(x,y) = [y + \frac{\hbar i}{m^* \omega} (\sin \frac{\omega x}{2} \sin \frac{\omega(t-x)}{2} / \sin \frac{\omega t}{2}]$$

Here the term in the square bracket is $G_o(0,0,t)$ while the exponential term is the first cumulant correction, aside from $e^{iEt/\hbar}$.

To obtain the limit of $\rho(E)$ valid in the tail region where E is large and negative we first take the $t \to \infty$ limit of the integrand in (5.10). This means only in the ground states, the lowest energy states, will be retained in $\rho_1(E)$. The integrals (5.10) cna then be perfomed exactly and, in the reduced unit $\nu = -(E - V_o)/E_Q$, the result is[9]

$$\rho_1(\nu, z) = \frac{Q^3}{E_Q\xi'^2} a(\nu, z) e^{-\frac{b(\nu, z)}{2\xi'}} \quad (5.11)$$

Here $E_Q = \hbar^2 Q^2/2m^*$ is again, as in the Halperin and Lax result (4.11), the energy associated with the impurity potential and all energies are expressed in units of E_Q. For example, $\nu = (V_o - E)/E_Q$ is, as in (4.11), the electron energy measured away from the mean potential V_o in units of E_Q and $z = (2E_Q/\hbar\omega)^{\frac{1}{2}}$ is a convenient reduced variational parameter, replacing ω. In the harmonic model the kinetic (zero point) energy of electron localization is just $3/4\ \hbar\omega$ and in reduced units is $T = (3/4)\hbar\omega/E_Q = 3/2\ z^{-2}$. The $a(\nu, z)$ and $b(\nu, z)$ are again the "universal", dimensionless functions but now we have relatively simple analytic expressions for them,[9]

$$a(\nu, z) = (T + \nu)^{3/2} / \left[8\pi\sqrt{2}\ z^6 \exp(\tfrac{z^2}{2}) D^2_{-3}(z) \right] \quad (5.12)$$

$$b(\nu, z) = (T + \nu)^2 \sqrt{\pi} / \left[2\sqrt{2}\ \exp(\tfrac{z^2}{4}) D_{-3}(z) \right] \quad (5.13)$$

where $D_p(z)$ is the parabolic cylinder function.

The subscript 1 or $\rho_1(E)$ means that the first order cumulant correction for the difference between W and the harmonic model has been evaluated exactly. Higher order cumulants which correspond approximately to the higher order corrections considered by Halperin and Lax have beeen ignored.

We may also obtain the semiclassical limit of $\rho_1^{SC}(E)$ from the full $\rho_1(E)$ in (5.10). This is obtained by letting $t \to 0$ in the integrand of (5.10) which corresponds to retaining only high electron energy states in $\rho_1(E)$. The integrals in (5.10) can again be performed giving

$$\rho_1^{SC}(E) = \frac{C}{2\sqrt{2}} \xi^{\frac{1}{4}} e^{-\frac{(E-V_o)^2}{4\xi}} D_{-3/2}\left(\frac{V_o-E}{\xi}\right) \qquad (5.14)$$

which is the result (3.16) derived by Kane[5]. In the limit $E-V_o \longrightarrow \infty$ this becomes

$$\rho_1^{SC}(E) = \begin{cases} \frac{C}{2\sqrt{2}\xi}(V_o-E)^{-3/2}\, e^{-\frac{(E-V_o)^2}{2\xi}}, & \frac{E-V_o}{\xi} \ll -1 \qquad (5.15) \\ C\sqrt{E-V_o}, & \frac{E-V_o}{\xi} \gg 1 \qquad (5.16) \end{cases}$$

Here (5.15) is Kane's well known band tail $\rho_1(E)$ while (5.16) is the free electron or parabolic band value valid for positive $E - V_o$ only. Note that (5.14) is independent of the variational parameter.

The density of states $\rho_1(E, z)$ of (5.14) in the band tail region (with the universal dimensionless functions $a(\nu, z)$ and $b(\nu, z)$ given by (5.12) and (5.13) clearly depends on the variational parameter z. The value of z determines the curvature of the optimum harmonic well which models $W(\vec{r} - \vec{r}')$. To evaluate $\rho_1(\nu, z)$ explicitly we need some variational principle to determine this optimum curvature z.

In 1975 Lloyd and Best [9] showed that the density of state $\rho_1(E)$ at energy E should maximize the pressure.

$$P(E, z) = \int_{-\infty}^{E} dE' \int_{-\infty}^{E'} dE'' \rho(E'', z) \tag{5.18}$$

of a hypothetical free Fermion system. This may be used here to determine the free parameter z in a band tail expression for $P(E, z)$. In terms of the reduced energy $\nu = (V_o - E)/E_Q$, this is

$$P(\nu, z) = -E_Q^2 \int_{\nu}^{\infty} (\nu - \nu')\rho(\nu', z)\ d\nu'. \tag{5.19}$$

On substituting (5.11) for $\rho(\nu', z)$ in (5.19) and maximizing $P(\nu, z)$ with respect to $z(dP(\nu, z)/dz = 0)$ we obtain the variational equation for z^1,

$$\int_{\nu}^{\infty} d\nu'(\nu\ \nu')a(\nu',z)e^{-b(\nu',z)/2\xi'}$$

$$x\left\{[+\frac{2D_{-4}(z)}{D_{-3}(z)} - \frac{1}{z}(\frac{T}{T+\nu'} + 2)] - \frac{b(\nu',z)}{2\xi'}[\frac{D_{-4}(z)}{D_{-3}(z)} - \frac{2z^{-3}}{(T+\nu')}]\right\} = 0$$

(5.20)

The $\rho_1(E)$ with z determined from (5.20) is denoted case 1. Since this is a lengthly expression we investigate two approximations. If we determine z by maximizing $\rho_1(E)$ itself, the equation for z is obtained by setting the curly bracket in (5.20) equal to zero.

$$[\frac{2D_{-4}(z)}{D_{-3}(z)} - \frac{1}{z}(\frac{T}{T+\nu} +2\ ] - \frac{b(\nu,z)}{2\xi'}[\frac{D_{-4}(z)}{D_{-3}(z)} - \frac{2z^{-3}}{(T+\nu)}] = 0 \qquad (5.21)$$

The $\rho_1(E)$ with z determined in this way is denoted as case 2. Finally, since $\rho_1(E)$ will be dominated in the low energy tail region by the exponent, we can follow Halperin and Lax and obtain z by minimizing $b(\nu, z)/2\xi'$. This corresponds to their equation (4.9) used to obtain the optimum wave function f. It is an approximation to maximizing, $\rho_1(E)$. Here the equation for z in this case (denoted case 3) is obtained by setting the last bracket in (5.20) equal to zero.

$$\left[\frac{D_{-4}(z)}{D_{-3}(z)} - \frac{2z^{-3}}{(T+\nu)}\right] = 0 \tag{5.22}$$

The expression (5.1) plus one of the three methods (case 1, case 2 or case 3) for determining z completely determines the density of state $\rho_1(E)$ in the band tail region. The $\rho_1(E)$ obtained in the shown in Fig. 7 for value of the dimensionless fluctuation parameter $\xi' = 50$.

To discuss the band tail density of states in Fig. 7 we note first that case 3, with z determined by minimizing the exponent only of $\rho_1(E)$, agrees numerically with Halperin and Lax's $\rho_1(E)$, who also minimized the exponent of $\rho_1(E)$. This provides an excellent point of contact with previous work. For this reason, case 3 is denoted the Halperin and Lax limit in Fig. 7. It is interesting to consider other quantities such as the exponent $n(\nu)$ defined by (4.13) and the kinetic energy of localization $T(\nu)$. Both quantities can be expressed a analytically as

$$n(\nu) = 3\nu/(T+\nu) \tag{5.23}$$

and

$$T(\nu) = \frac{3}{2}z^{-2} \tag{5.24}$$

The asymptotic values of $a(\nu)$, $b(\nu)$, $n(\nu)$ and $T(\nu)$ are given in Table 1.

Next we note that case 1 and 2 are almost identical. This means that $\rho_1(E)$ with z obtained by minimizing $\rho_1(E)$ itself is almost identical to $\rho_1(E)$ obtained using the rigorous Lloyd-Best variational principle. It is only case 1 with z determined using the Lloyd-Best variational principle that has a rigorous foundation. However, case 2 is clearly a good approximation and we denote case 1 and 2 as the present theory. The case 3, however, is not such a good approximation and particularly shows an unphysical down turn in $\rho_1(E)$ in the range $\nu \sim 0$ to 5.

The band tail limit of $\rho_1(E)$ in (5.11) is valid only for low (negative) energy E in the band tail region (large = $\nu = -(E - V_o)/E_Q$). Since $\rho_1(E)$ crosses the Kane theory, $\rho^{SC}(E)$, which is valid at higher E, an interpolation scheme is suggested. At low E we should use the band tail limit $\rho_1(E)$ given by (5.11) with z determined from (5.20) up to the point where $\rho_1(E)$ crosses $\rho^{SC}(E)$. Thereafter higher energy use Kane's $\rho^{SC}(E)$ in (5.15) This provides a density of states over the whole energy range.

The optical absorption $\alpha(\omega)$ in heavily doped n-and p-type GaAs is evaluated using the interpolation density of states to test the sensitivity of $\alpha(\omega)$ on $\rho(E)$.[22]

VI Discussion

In this paper we have discussed two main effects, the shift of band edge and the band tailing of heavily doped semiconductors. The shift of band edge can be sucessfully determined and shown to be arising from the exchange effect.[1]

The band tailing effect arised from the random arrangement of impurities is a statistical problem depending on the configuration averaging. The density of states calculated in the perturbation technique, which neglected the potential fluctuation, show a sharp cut-off in the tail region.

In Kane theory, the electron is required to respond to their fluctuation. This problem is then reduced to finding the function describing these fluctuation which turns out to be Gaussian. The tail found by Kane is also Gaussian which overestimate the tail s states. This failure aries from the use of semiclassical model which neglects the kinetic energy of localization or zero point energies. This theory is reasonable for high energies states but not for deep states. However, this method is the simplest description of the band shape in disordered semiconductors which give rough approximate value of $\rho(E)$ through out the band.

The quantum mechanical model of Halperin and Lax which include the kinetic energy of localization is more precise than the semiclassical theory of Kane. The calculation leads to the density of states $\rho(E) \sim e^{-E^n}$ where n vary from $\frac{1}{2}$ and 2. Because they did not obtain the analytical expression for $\rho(E)$, it is inconvenient to apply their result to any specific calculation.

The the path integral method the $\rho(E)$ can be expressed in the analytical terms which allow one to study several dimensionalless function $a(\nu)$, $b(\nu)$, $n(\nu)$ and $T(\nu)$ asymptotically. Furthermore, the calculation can be done to satisfy the Lloyd and Best variational principle which gives a substantial improvement over the original work of Halperin and Lax and of Kane. This improved band tail is used to obtain the $\rho(E)$ for all energies.

Acknowledgement

This work was supproted by the UNIT CELL program of Chulalongkorn University and the International Seminar in Physics, Uppsala, Sweden.

References

1. Shklorskii, B. I., Efron, A. L., *Electronic Properties of Doped Semiconductor*, Springer-Verlag, 1984.
2. Fistul, V. *Heavily Doped Semiconductors*, New York: Plenum Press, 1969.
3. Wallace, P. R., Harris, R., Zuckermann, M. J., eds. *New Develops in Semiconductors*, Leyden: Noordhoff International Pub., 1973.
4. Efros, A. L., *Sov. Phys. Usp.*, vol. 16, p. 789, 1974.
5. Kane, O. E., *Phys. Rev.*, vol. 131, p. 79, 1963.
6. Halperin, B. I., Lax, M., *Phys. Rev.*, vol. 148, p. 722, 1966.
7. Halperin, B. I., Lax, M., *Phys. Rev.*, vol. 153, p. 802, 1967.
8. Sa-yakanit, V. J., *Phys. C: Solid State Phys*, vol. 7, p. 2849, 1974.
9. Sa-yakanit, V., *Phys. Rev. B*, vol. 19, p. 2266, 1979.
10. Sa-yakanit, V., Glyde, H. R., *Phys. Rev. B*, vol. 22, p. 6222, 1980.
11. Feynman, R. P., Hibbs, A. R., *Quantum Mechanics and Path Integral*, New York: McGraw-Hill Book Co., 1965.
12. Edwards, D. F., Gulyaev, Y. B., *Proc. Phys. Soc.*, vol. 83, p. 495, 1964.
13. Edwards, S. F., *Adv. Phys.*, vol. 16, p. 359, 1967.
14. Jones, R., Lukes, T., *Proc. Roy. Soc. Lond.*, vol. A309, p. 457, 1969.
15. Bezak, V. *Proc. R. Soc. Lond.*, vol. A315, p. 339, 2970, *J. Phys. A*, vol. 4, p. 324, 1971.
16. Papadopoulos, G. J., *J. Phys. A*, Vol. 7, p. 183, 1974.
17. Friedberg, R., Luttinger, J. M., *Phys. Rev. B.*, vol. 12, p. 4460, 1975.
18. Freed, K. F., *J. Phys. C*, vol. 4, p. L331, 1971; Freed, K. F., *Phys. Rev. B*, vol. 5, p. 4802, 1972.
19. Lloyd, P., Best, P. R., *J. Phys. C*, vol. 3, p. 3752, 1975.
20. Kane, E. O., *J. Phys. Chem. Solids*, vol. 1, p. 249, 1957.
21. Eagles, D. W., *J. Phys. Chem. Solids*, vol. 16, p. 76, 1960.
22. Stritrakool, W., Sa-yakanit, V., Glyde, H. R., *Phys. Rev. B*, vol. 32, 1090, 1985.

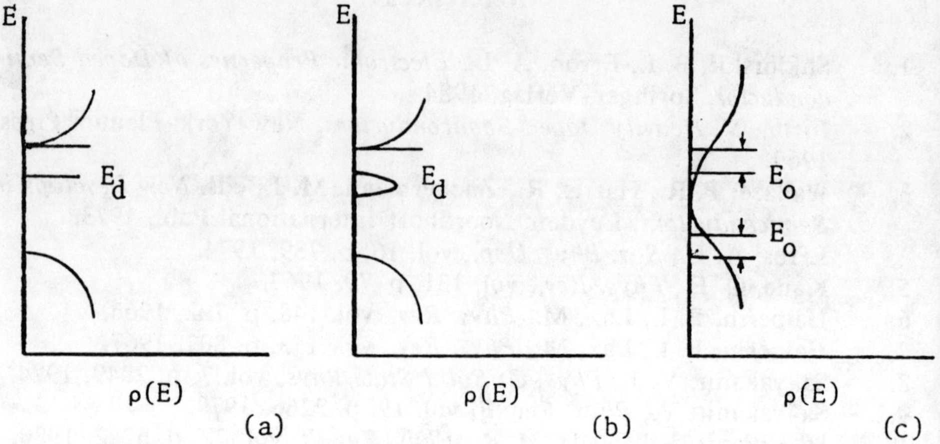

Fig. 1. The density of states $\rho(E)$ as a function of energy E. (a) Single impurity atom; (b) Low impurity concentration localized impurity level located at an energy level E_d below the conduction band; and (c) Heavily doped semiconductor-merged impurity band showing the band tails and the shift of the band edges E_0.

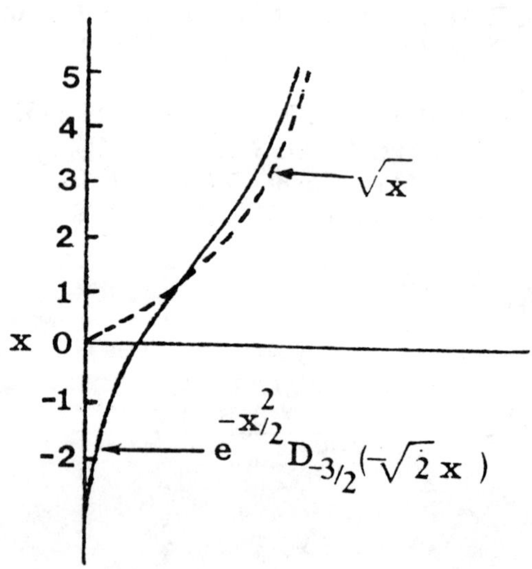

Fig. 2. The free particle density of states and the Kane function.

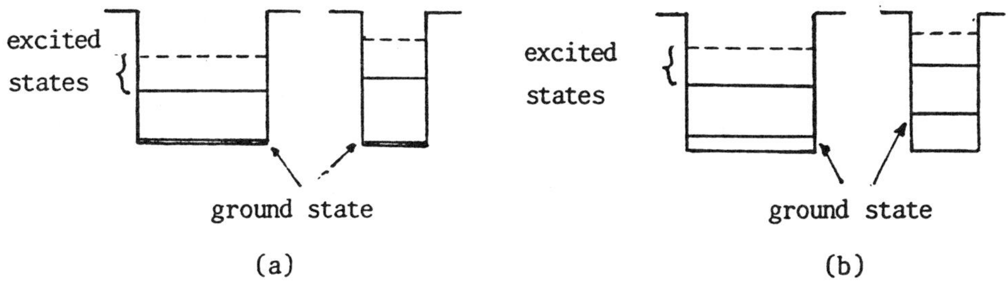

Fig. 3. (a) Semiclassical energy levels, (b) Energy levels obtained using quantum theory including zero point energy.

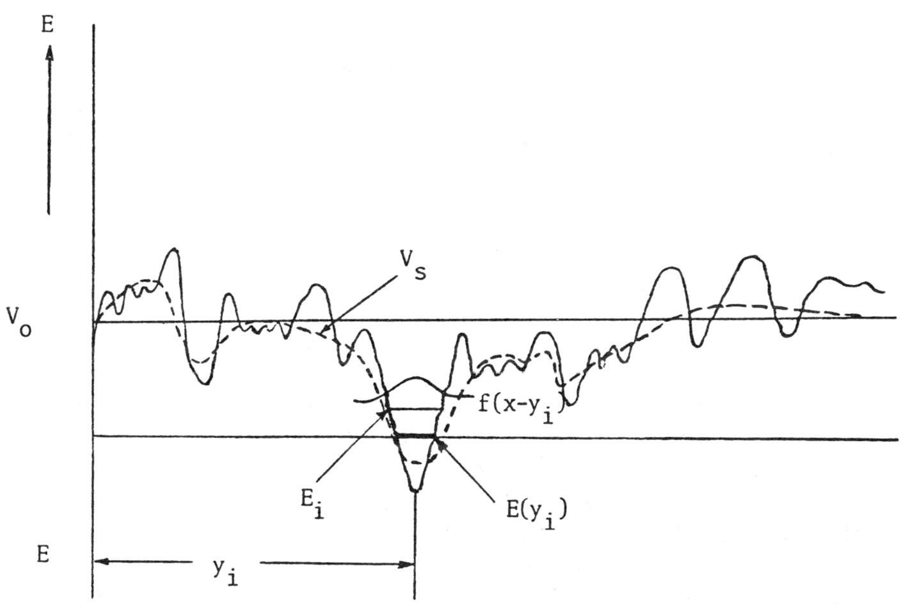

Fig. 4. The random potential, smoothed potential V_s and the trial wave function $f(x - y_i)$.

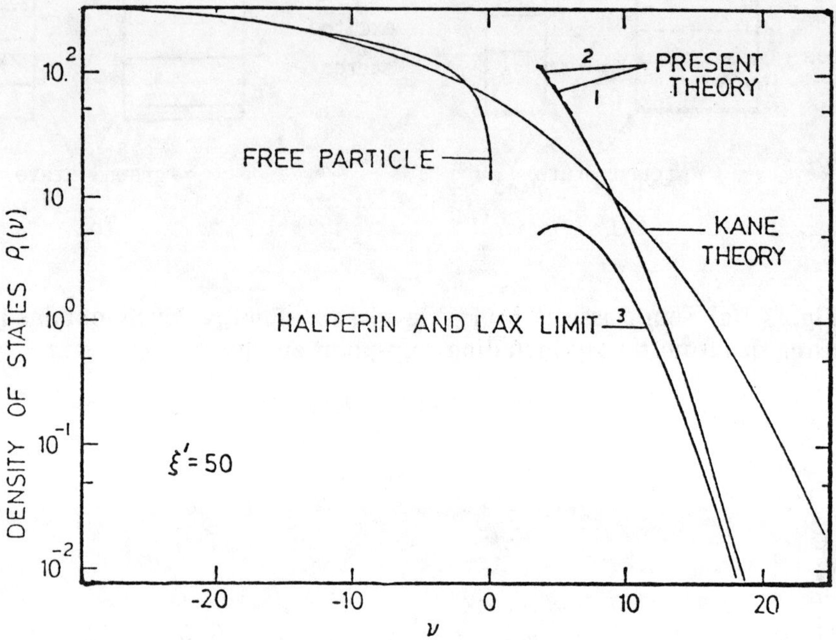

Fig. 5. The density of states

$$\rho_1(v, z) = \frac{Q^3}{E_Q \xi'^2} a(v, z) e^{-\frac{b(v, z)}{2\xi'}}$$

in units of $(Q^3/E_Q\xi'^2)$ for $\xi' = 8\pi m^{*2} z^2 e^4 n / Q^5 \hbar^4 \varepsilon_0^2 = 50$.

The present theory is the band-tail value $\rho_1(E)$ with the variational parameter z determined by (Case 1) the Lloyd-Best variational principle (Equation (5.20)), and by (Case 2) maximizing $\rho_1(E)$ only (Equation (5.21)). Case 1 and 2 are indistinguishable at $\xi' = 50$. The **Halperin and Lax limit** is $\rho_1(E)$ with z determined by maximizing the exponent in $\rho_1(E)$ (Case 3) (Equation (5.22)). The **Kane and Free Particle** values are Kane's semiclassical result (3.16) and (3.15), respectively.

HOT TOPICS IN SEMICONDUCTOR PHYSICS AN OVERVIEW

M. Zafar Iqbal
Semiconductor Physics Laboratory
Department of Physics, Quaid-i-Azam University
Islamabad (Pakistan)

Abstract: **An overview of the current research activity in semiconductor physics is presented. Some of the exciting new developments will be highlighted after a brief background discussion. Some of the major controversies leading to directions for future research will be discussed. Areas reviewed include quantum wells, superlattices, and two-dimensional electron gas, lattice defects, amorphous semiconductors, and scanning tunneling microscopy of semiconductors.**

These two presentations are intended to convey the flavor of the current excitement in the field of semiconductor physics by presenting an overview of the various subfields in which intense research activity is concentrated. These areas include:

- quantum wells, superlattices, and two-dimensional electron gas,
- defects in semiconductors,
- amorphous semiconductors,
- scanning tunneling microscopy.

it is perhaps pertinent to point out for the benefit of the generalist, that a persistent undercurrent of the material presented here will be the following feedback loop:

"Pure" Science

Technology

"pure" science generates new technology thus making the implementation of new ideas and novel devices possible which in turn leads to the unravelling of hitherto unknown physical phenomena pushing further the frontiers of physics. Although, in general true about all science, this interface is perhaps nowhere more intimate and obviated by a rapid mutual transfer between science and technology than in semiconductor physics as exemplified by the many developments discussed in these lectures.

1. Quantum Wells, Superlattices, and Two-Dimensional Electron Gas

It is probably no exaggeration to say that the greatest research effort in semiconductor physics is currently concentrated in this area, motivated both by the exciting basic physics and immense potential for novel device applications. We shall discuss some of these novel devices later. The new phenomena of quantum Hall effect and the fractional quantum Hall effect will, however, not be dealt with here since Dr. Gallagher's presentation deals with this subject in detail.

1.1. Basic Idea

The basic physics of these systems revolves around the simple idea that carrier motion is quantized when it is confined in an ultra-thin (~100 Å) semiconductor layer. The theoretical details of this problem have been discussed by Dr. Berggren at this symposium. This new situation gives rise to a system of electrons and holes free to move in plane of the layer but confined in the transverse direction, a 2-D electron gas leads to novel transport properties and fundamentally new effects, not present in the bulk. For instance, calculations by Tsu and Esaki [1] (1973) predicted the phenomenon of "resonant tunneling" of electrons through quantum wells with enhanced transmission coefficient (T^*T) whenever the Fermi-level in the barrier (or electrode) resonates with the discrete quantum well states (Figure 1). This was experimentally confirmed soon thereafter [2] although a fuller confirmation of the associated negative differential resistance effects and their potential had to wait until recently for perfection of the fabrication technology.

1.2. Superlattice

Figure 1 shows splitting of the quantum well states into narrowly spaced groups of states as the number of closely spaced (50 Å or less) wells (or their separating barriers) is increased. This is reminiscent of the splitting of the discrete atomic states on bringing a number of atoms together at small but regular intervals to form a crystal lattice, leading to the formation of energy bands represented in the various Brillouin zones in the large number of atoms limit. Figure 1 indicates that in the large number of wells, $n \to \infty$ limit, we have the formation of minibands consisting of groups of very closely spaced energy levels, rather than the discrete quantum well states of a single quantum well. In the ideal case of strictly two-dimensional wells the situation corresponds to a one-dimensional "Kronig-Penney Crystal" with the periodicity of the potential-well repetition. Such a structure of periodically repeated potential wells is known as a "superlattice" superposed on the ordinary lattice of atomic dimensions. This system has its own mini-Brillouin zones and gaps in the E-k dispersion relation in the superlattice direction. The ability to tailor the width and depth of 2-D quantum wells in such a structure opens up the possibility of manipulating the shape of the overall one-dimensional potential superposed on the system of electrons providing opportunities for interesting "do-it-yourself" quantum mechanics as well as a tremendous potential for novel applications.

Two types of superlattices can be produced using molecular beam epitaxy (MBE) and metal organic chemical vapor deposition (MOCVD) techniques:

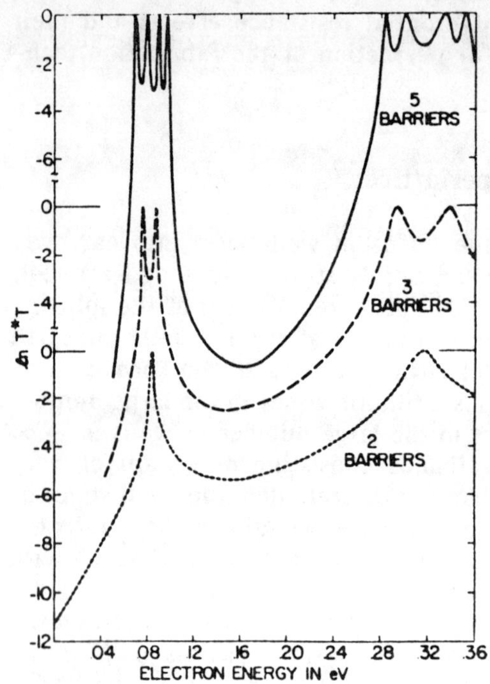

Fig. 1. Transmission coefficient vs. the electron energy in eV for the cases of a double-, triple-, and quintuple-barrier. The well widths are 20 and 50 Å, respectively. The barrier height is 0.5 eV (Tsu and Esaki 1973).

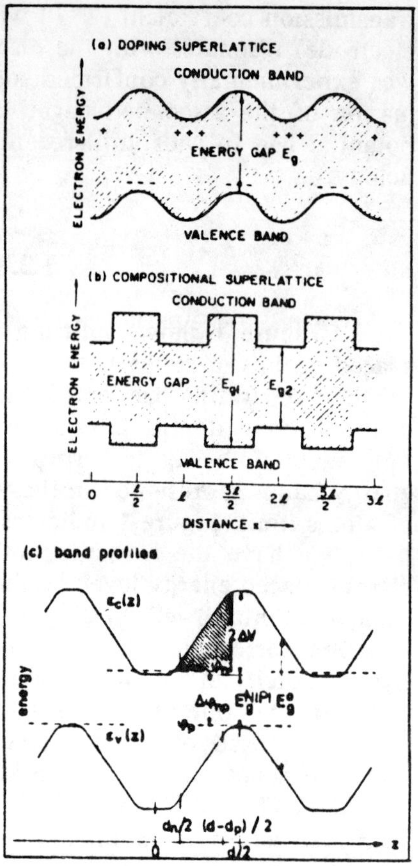

Fig. 2. Spatial variation of the conduction and valence gand edges (a) doping, (b) compositional, and (c) nipi superlattices.

a. Compositional Superlattices

In this type of superlattice the barriers and wells are produced by alternating the composition of the material between two different bandgaps. The archetypal example of this class is the GaAs-$Al_xGa_{1-x}As$ superlattice.

These consist of GaAs quantum well layers enclosed between barriers of the higher bandgap $Al_xGa_{1-x}As$ ($x < 0.4$) both of which are direct gap, almost perfectly lattice matched semiconductors. The bandgap discontinuity, still a matter of active theoretical research, is primarily accommodated in the conduction band in these structures with the electron and hole quantum wells occupying the same spatial position as shown in Figure 21.

b. Doping or "Nipi" Superlattices.

Here the superlattice structure is created by alternating the nature of the dopants in consecutive layers of the same material using the fine control over doping level and layer thickness possible with MBE growth, resulting in the band structure shown in Figure 2b. The n-type electron well regions are thus interleaved by p-type barrier regions leading to a spatial separation of the electron and hole quantum wells or an indirect gap in real space. This effect, which leads to special properties of the structure, is further enhanced by interposing thin regions of the intrinsic material often referred to as the "nipi" structure (Figure 2c).

The optical absorption studies [3] on compositional multiquantum well structures revealed spectral features (Figure 3) in perfect agreement with the predicted transitions between the electron and hold quantum well states providing direct evidence of the quantization of the 2-D electron gas. In addition such studies revealed exciton formation between bound state electrons and holes in the quantum wells as well as finer features accountable in terms of separate series of quantum well states for the light and heavy hole bands in GaAs as seen in Figure 3.

1.1.3. Technological Spin-Offs

In the following we will point out some of the prominent applications emerging with the development of the quantum well structures.

1. Modulation Doping and High Electron Mobility Transfer. Modulation doping, i.e., incorporation of dopant atoms (such as Si) exclusively in the barrier regions (i.e., AlGaAs) of a multiquantum well structure feasible with the MBE technique has mad possible semiconductor (GaAs) material with hitherto unapproachable quality. This arises because of the ability of the quantum well sandwich layer to accumulate all the donated electrons from the high energy barrier region as shown in Figure 4. These electrons

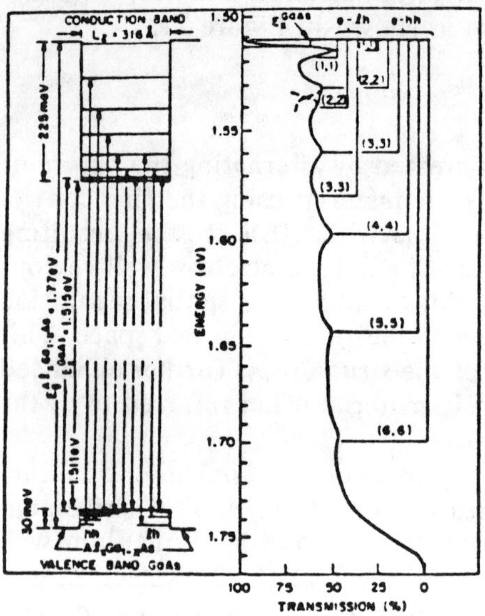

Fig. 3. Transmission spectrum of a heterostructure of 316 Å thick GaAs layers and 250 Å thick $Al_{0.21}Ga_{0.79}As$ barriers. The predicted energy levels and transition energies corrected for exciton effects are shown in rectangular potential wells, $\Delta E_{CB} = 0.88\ \Delta E_G$ (R. Dingle 1975).

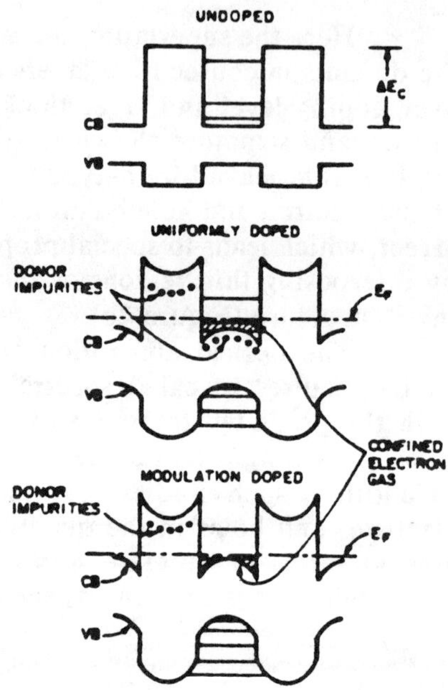

Fig. 4. Energy band diagram for *n*-doped and undoped GaAs-$Al_xGa_{1-x}As$ superlattices.

moving in the relatively dopant-free well region can consequently reach mobilities as high as 3,000,000 $Cm^2/V\text{-}sec^4$, which were previously impossible for bulk GaAs.

This property found immediate application in the form of an ultrafast field effect transistor structure variously termed TEGFET (two-dimensional electron gas FET), SDHT (selectively doped heterostructure transistor), or simply HEMT (high electron mobility transistor) with switching times reaching 5.8 ps [5].

2. Bandgap/Wavefunction Engineering — Novel Optical Devices. With the precise control afforded by current computer-controlled MBE facilities, the composition of III-IV compound semiconductors, and with that the bandgap, can be graded both abruptly and gradually. This allows immense possibilities for engineering the electron wavefunctions by tailoring the overall form of the 1-D potential. New artificially structured materials and novel devices can be created. These applications, two of which are briefly discussed below, are having a tremendous impact on the entire photonics (optical devices) industry.

a. Avalanche photodiode (APD) and solid state photomultiplier. In 1983 a superlattice avalanche photodiode of GaAs-AlGaAs was demonstrated to have a much higher ratio of electron and hole ionization coefficients leading to a higher gain and greatly suppressed noise [6], in contrast with the corresponding values for bulk GaAs where this ratio is almost one. This enhancement arises from the high asymmetry between the magnitudes of the band offsets for electrons and holes at the GaAs-AlGaAs junctions.

A staircase APD where the superlattice material is graded in terms of bandgap from one end to the other as shown in Figure 5 is expected to provide a truly solid-state photomultiplier for long-wavelength optical communication systems. Here avalanche ionization, almost exclusively restricted to electrons, is almost entirely sustained by large band offsets localized at band discontinuities which in turn implies extremely low gain fluctuations.

b. Quantum well lasers. Carrier confinement in quantum wells leads to a lowering of threshold currents for laser action. Bandgap engineering can be exploited to grade the refractive index, thereby making it possible to confine injected carriers and light in separate regions. This means very low threshold current values for laser action have been achieved.

Further, using multiquantum well lasers, the laser wavelength can be tuned by varying the thickness of the quantum well region. Using InGaAs wells and AlInAs barriers in such a structure, for example, laser wavelengths from 1.5 μm to 1.7 μm have been achieved by varying the well width form 80 Å to 1000 Å (Figure 6) [7].

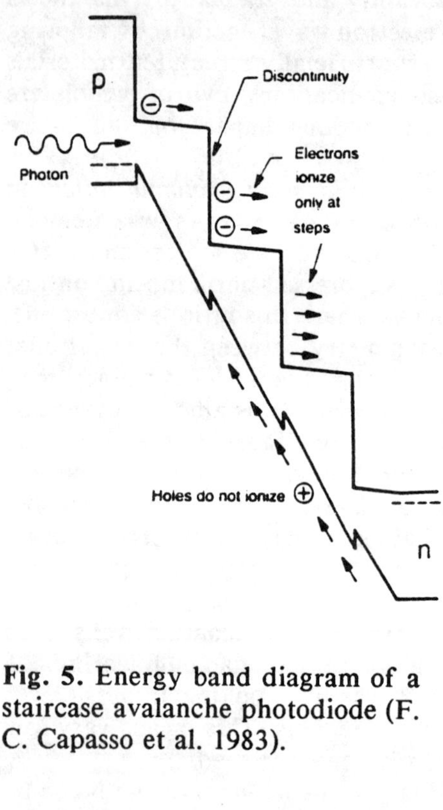

Fig. 5. Energy band diagram of a staircase avalanche photodiode (F. C. Capasso et al. 1983).

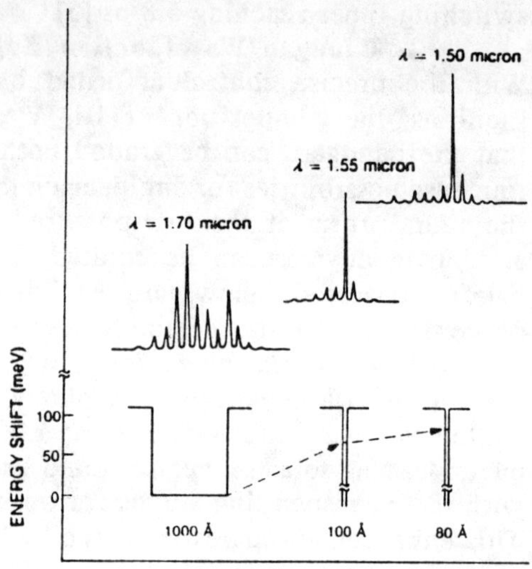

Fig. 6. Long-wavelength multi-quantum-well laser with $Ga_{0.47}In_{0.53}As$ wells and $Al_{0.48}In_{0.52}As$ barriers. By adjusting the well width from 1000 Å to 80 Å, the dominant mode spectrum of the laser can be tuned from 1.7 μ to 1.5 μ (Temkin et al. 1983).

2. Defects in Semiconductors

Defect studies have traditionally been at the heart of semiconductor materials and devices research, in the process providing a wealth of interesting and often exciting basic physics insight. We shall provide a flavor of this excitement by discussing some of the hotly pursued specific systems in this area.

2.1. EL2 in GaAs

This is a subject of intense activity due to the obvious technological promise of GaAs.

EL2 is the dominant midgap (thermal activation energy 0.815 eV) level native defect in Czochralski, Bridgman and MOVPE (metal organic vapor phase epitaxy) grown GaAs. It is responsible for the semi-insulating behavior of Czockralski and Bridgman GaAs due to compensation of the residual acceptors — a factor largely responsible for its high technological importance. But research on this deep level defect is also motivated by the stimulating unusual physical properties exhibited by it. Perhaps the single most important property of EL2, and one that has until recently been regarded as its most distinct fingerprint, is the photoquenching or bleaching effect.

This property, manifested in many different optical experiments, can be illustrated in the photocapacitance experiments first demonstrated by Vincent and Bois [8]. Starting with EL2 deep levels filled with electrons in an *n*-type GaAs Schottky barrier or *pn* junction device, if a light of energy $0.75\ \text{eV} < h\nu < E_G$ is shown we obtain a monotonically rising photocapacitance signal due to electron emission into the conduction band in the usual manner, as shown by curve (a) in Figure 7. This happens as long as the sample temperature is between 100 K and 250 K. If, however, the same experiment is carried out below 100 K, after an initial rise the capacitance shows a decline rather than saturating, curve (b); finally approaching the initial value with a characteristic decay time which is about 100 times longer than the rise time of curve (a). No amount of illumination can subsequently restore curve (a). Recovery of curve (a) is only possible by either (a) heating to $T > 140$ K with an activation energy ~0.36 eV, or (b) injection of minority carriers at $T > 30$ K — this recovery process is also thermally activated with an activation energy ~0.106 eV.

This phenomenon, also observed in optical absorption and photoluminescence experiments, is interpreted in terms of a transformation of EL2 from the ground state $EL2^0$ to some metastable excited state EL2* which is inaccessible to illumination.

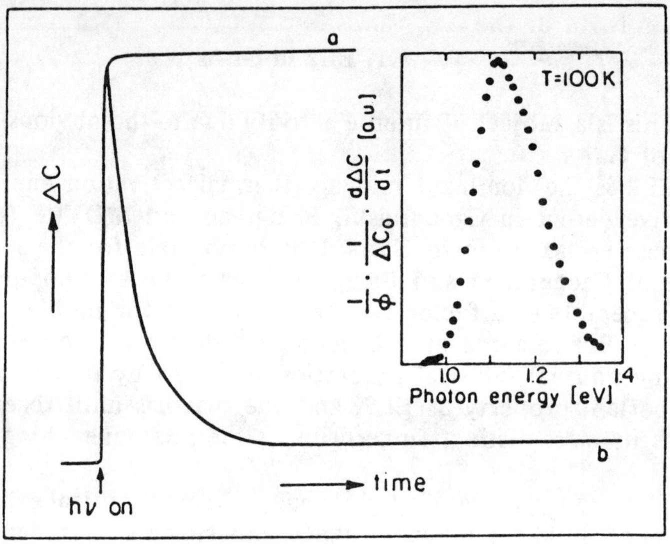

Fig. 7. Photocapacitance transients observed with EL2 in GaAs: a) sample temperature between 150 and 250 K, showing the fast transient, b) sample temperature lower than 100 K, showing the slow transient from the photocapacitance quenching effect (Vincent and Bois 1978). The inset contains the spectral distribution of the photoquenching rate.

The microscopic identification of EL2 has been a subject of sizzling controversy over the last nearly ten years and continues to attract intense research effort. This state of affairs is perhaps best summed up by the remark of Professor Gatos at the 1984 International Conference on Defects in Semiconductors (ICDS): "The questions seem to outnumber the answers."

After an initial erroneous identification with oxygen which produces an electron level close to the EL2 energy level, the consensus seemed to develop in favor of the As_{Ga} antisite defect. As_{Ga} appeared to be a strong candidate for EL2 on the basis of:

1. stoichiometry dependence of EL2, i.e., abundance of EL2 in material grown in As rich conditions;
2. photo-EPR studies pointing to a midgap donor level originating from As_{Ga};
3. photoquenching properties observed for the EPR signal from the As_{Ga} surrounded by four nearest neighbor As atoms (As_{Ga}-As):
4. Uniaxial stress studies of the fine structure of optical absorption with zero phonon line (ZPL) at 1.039 eV strongly supports a neutral, undistorted antisite defect.

However, the initial euphoria about the antisite identification soon subsided because of the following problems:

1. the intensity of 1.039 eV ZPL does not correlate with EL2 concentration;
2. recent magnetic circular dichroism observations contradict As_{Ga} identification;
3. metastability of EL2 is difficult to explain in terms of an isolated As_{Ga} defect since such a well bonded (to 4 As nearest neighbors) defect could not undergo the large lattice relaxations that appear to be necessary for metastability. This necessitates additional defects forming complexes with the As_{Ga} antisite.

As a result the As_{Ga}-X complex, where X is the as yet unidentified partner, is believed to be the most probable origin of EL2. The current favorite for X is the interstitial Arsenic atom (As_i), although a number of other candidates are on the table. The As_{Ga}-As_i defect is supported by the recent combined DLTS and EPR studies as well as the ENDOR (electron nuclear double resonance) work. The recent ENDOR studies have, however, revealed a family of antisite complex defects raising the possibility of a family of EL2 centers. The question of the microscopic origin of EL2 is thus far from settled, and continues to be a subject of active research.

2.1. Thermal Donors in Si

A complete understanding of the "thermal donors" known to be produced by prolonged 450°C heat treatment of Czochralski grown silicon crystals has been a challenge for semiconductor physicists in the 32 years since their discovery. A series of nine double donors TD1-TD9 with thermal activation energies of 69.9 Mev to 53.0 MeV, have been detected, each of which displays effective-mass-like excited states in infrared absorption.

Although a lot of progress has recently been made in the modeling of thermal defects using deep level transient spectroscopy (DLTS), electron paramagnetic resonance (EPR), and high resolution electron microscopy (HREM), forcing Corbett to record "an almost euphoric excitement at Coronado that the thermal donor problem was almost solved" in the ICDS (1984) proceedings, understanding is far from complete.

2.2.1. Microscopic Structure

As a result of these studies a cluster of oxygen atoms is generally believed to be involved. The electronic structure seems to be consistent with a repulsive inner core that grows with the addition of oxygen atoms showing signs of electrical activity at some critical cluster size. Various models proposed for the core structure include Si-O-Si-O-Si chains along ⟨110⟩, Ylid model with its central instertial 0 atom core, and a central interstitial Si atom forming the active core (OBS model). All these, nevertheless, suffer from the snag that the required theoretical diffusion rate for oxygen is much larger than observed.

2.2.2. Recent New Observations

Recent observations include the bistability of TD1 predicted on the basis of proposed negative *U* behavior much earlier [9] which has recently been observed [10] for the first time in infrared studies. A second new development is the recent observation of a new NL10 line in EPR spectrum in addition to the usual NLB line by the Amsterdam group [11]. The same group has detected the hyperfine structure of ^{17}O in TDs demonstrating the involvement of oxygen in a direct experiment for the first time. The full impact of these new observations has yet to be realized.

Also little understood at the present time are the class of seven shallower (34.7 MeV to 37.4 MeV) thermal donors formed during the early stages (first 15 minimum) of 450°C annealing and the class of so-called "new thermal donors" (~0.17 eV) produced on prolonged annealing at 650°C.

Thus the qualified "euphoria" noted by Corbett notwithstanding, the current status of the problem is perhaps realistically reflected by the comments of Watkins at the ICDS (1986) two years later: "The many new pieces to the puzzle that have surfaced in the past two years are both exciting and bewildering. Clearly we may still have quite a way to go before we really understand thermal donors."

2.3. DX Centers (in $Al_xGa_{1-x}As$)

The DX defect center — detected in more than one III-IV compound semiconductor — is of great interest as much for its unusual properties as for its technological importance, particularly in $Al_xGa_{1-x}As$.

It is recognized as a deep donor found in an $Al_xGa_{1-x}As$ alloy semiconductor with the aluminum fraction $x > 0.26$ on incorporation of any of the commonly used donor dopants such as Si, Te, Se, Sn. It seems to be omnipresent in large concentrations in material grown by LPE, MOVPE, and MBE. Its obvious association with donor impurity doping led to its christening as DX center, i.e., a donor (D)-unknown defect(X) complex. Its presence, usually in large concentrations, gives rise to the following macroscopic effects:

- high material resistivity,
- persistent photoconductivity,
- saturable pulsations in lasers,
- collapse of I-V curves in GaAs-AlGaAs HEMT devices;

underscoring the obvious importance of the current flurry of interest in the problem.

2.3.1. Unusual Properties

The high technological interest in DX center is matched only by its fundamental physics aspcect brought to light by its unusual physical characteristics. These include:

- optical ionization energy larger than the thermal ionization energy (in the region of ~0.2 eV);
- thermally activated capture cross-section with a large activation energy (~0.3 eV) — a property leading to the observation of metastability and persistent photoconductivity at low temperatures;
- the DX deep donor level is pinned to the indirect (L-minimum) of the conduction band (Figure 8).

2.3.2. Models

Lang and Logan [12] first invoked the phenomenon of large lattice relaxation to explain the metastability of the DX center. According to their model, represented by the configuration coordinate diagram of Figure 9, the ionized level of the defect is resonant with the conduction band. On electron capture the configuration of the center changes pulling the level into the gap (for material with an Al fraction larger than ~0.2) as a consequence of the lattice relaxation. This nicely explains the unusual properties of the DX centers as indicated in Figure 9, where the energies required for various transitions are shown.

Coming in the wake of large lattice relaxation models of silicon vacancy and isolated oxygen center in GaP, this success of the model gave an air of universality to the large lattice relaxation in the approach to the understanding of metastability of defects in semiconductors, which appears to be a more common phenomenon than was first believed.

Before this belief in universal connection between metastability and large lattice relaxation could finally settle, however, a fresh look at the problem has been necessitated by the recent new observations on DX center made by Henning and Ansems [13]. These observations of new near-band-edge luminescence suggest an optical DX donor energy $E_D^{opt} \sim 200 \pm 20$ MeV with a thermal energy $E_D^{th} \sim 180$-200 MeV inferred from the temperature dependence of the intensity of the new luminescence. The spectral features of this new photoluminescence consist of a zero-phonon line and two LO-phonon replicas.

The strikingly close values of E_D^{th} and E_D^{opt} strongly suggest an inappreciable lattice relaxation in direct conflict with the model of Lang and Logan. In fact Henning and Ansems have put forward an alternative configuration coordinate diagram model shown in Figure 10 requiring small lattice relaxation. This model provides very good qualitative and quantitative explanations of most properties of the DX center. The model predicts an optical threshold for persistent photoconductivity close to 225 MeV in agreement with their observation of an experimental threshold at less than 320 MeV which, however, is in conflict with previous experiments.

The important implication of this model, if proved correct, lies in the fact that it brings into focus a purely electronic mechanism for metastability — a revival of the pre-large-lattice-relaxation era.

2.4. Hydrogen Induced Defect Passivation

Hydrogen has emerged as a focus of both experimental and theoretical research in recent defect studies. This stems from the ability of the

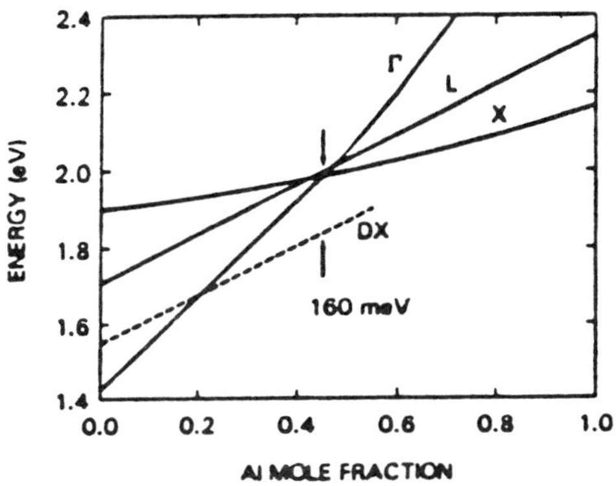

Fig. 8. Lower conduction band edges in $Al_xGa_{1-x}As$ as a function of x. Hall measurements indicate that the trap level associated with the *DX* center follows the *L* band edge and becomes resonant with the band for $x < 0.22$.

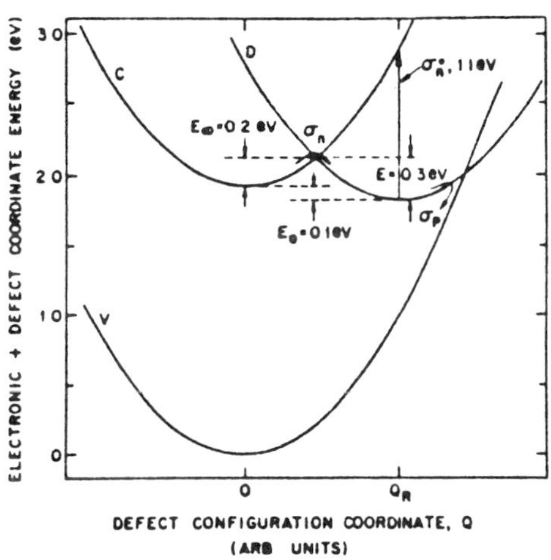

Fig. 9. Lang and Logan configuration coordinate model of *DX*.

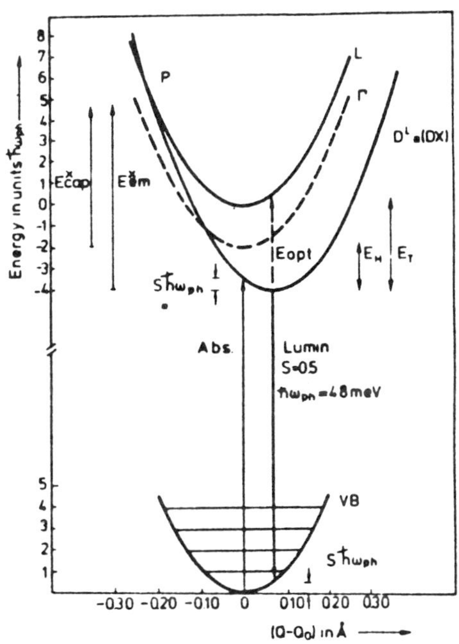

Fig. 10. Henning and Ansems configuration coordinate model of *DX*.

hydrogen defect to electrically neutralize (passivate) active defect centers in silicon as well as in some other semiconductors. These defects in Si include Au, Pd, Pt, Cu, Ni, Ag, Fe, O-V complex, and V-V complex. Passivation has mostly been observed in the case of acceptors in p-type material, although Johnson [14] recently found partial passivation of n-Si for the first time. The involvement of hydrogen in passivation has been demonstrated directly by Pankov [15] and lately by Johnson [16] who found the H profile to replicate the acceptor profile in p-Si, implying some kind of hydrogen-impurity pairing.

2.4.1. Model

Although H-impurity pairing is generally believed in, the nature of the electrical activity of hydrogen is currently a subject of strong controversy. Thus, the possibilities of a neutral hydrogen for all Fermi-level positions, an H (donor) state and an H (acceptor) state are hotly debated.

Recently Pantelides [17], in considering the six possible ways in which interstitial hydrogen could behave, concluded that only hydrogen deep donor can lead to direct compensation. This would imply pairing resulting from capture of free holes by hydrogen rather than vice-versa. However, in n-Si his model predicts direct pairing. But these conclusions are far from proven and await future experiments.

3. Amorphous Semiconductors

Although marred by the absence of longrange order, and defying understanding in terms of the general principles and concepts found so useful for the understanding of crystalline semiconductors; the field of amorphous semiconductors has made rapid progress due to its impressive technological potential. However, the field has been described as still in "early adolescence" by Fritzsche [18].

3.1. Amorphous Superlattices

After the fruitful early studies of crystalline superlattices, such structures seemed both attractive and feasible particularly in view of the freedom from lattice matching constraints. Structures with alternating layers of a-Si:H have been fabricated using plasma-deposition. In the later type of structures, the interesting phenomenon of "charge-transfer doping" has been observed and used to dope hydrogenated a-Si. This occurs due to transfer of electrons to or from Si and the widegap SiO_x or SiN_x prompted by the work function difference. Accordingly, large *n*- or *p*-type *a*-Si:H conductivities can be achieved.

Of particular interest are the doped amorphous superlattices theoretically investigated by Döhler [19]. Here the ability to continuously engineer the bandgap by optical band-to-band excitation or by injection and the indirect bandgap in real space are extremely attractive features. The latter results in extremely long recombination lifetimes called the photoinduced charge-storage effect by Döhler. The study of the amorphous doping superlattice is particularly rewarding insofar as it leads to knowledge and understanding of the amorphous bulk materials such as the density of gap states which is a very important characteristic of an amorphous semiconductor. The optical charge storage effect has seemingly been observed in an *a*-Si:H doping superlattice even at room temperature [20].

3.2. Density of Gap States, Transport Properties

The subject of the density of states (DOS) in the pseudogap of amorphous silicon is still unsettled both experimentally and theoretically. The results obtained by various experimental techniques do not agree with each other.

The electrical transport properties continue to cause surprises and are replete with unsolved problems. These include the exact nature of the conduction mechanism, the recently observed conductivity kinks in log vs. 1/T plot, the (opposite) sign of the Hall effect, and the Staebler-Wronski effect whereby light is believed to induce metastable defects thought to be dangling bands produced during photogenerated electron-hole recombination.

Applications Research. In spite of a lack of understanding for a host of basic properties, the field is already rich with applications providing further drive for research. These areas of rapid progress include:

a. solar energy conversion devices - large scale tandem arrangements of optically complementary alloys tailored to be sensitive to different parts of the solar spectrum are expected to boost the efficiency of large area *a*-Si solar cells;

b. large area FETs or PIN diodes of thin *a*-Si films are being developed for isolation and address circuits in large area flat-panel liquid crystal displays;

c. amorphous superlattices have found immediate application as focusing and dispersive (grating) devices in UV and X-ray optics;

d. xerographic photoreceptors of *a*-Si have been developed to provide high copying speed and longer life to photocopy machines.

4. Scanning Tunneling Microscopy

The recent development of scanning tunneling microscopy has brought out rare insight into the structure of semiconductor surfaces. This has been made possible by the unprecedented resolution afforded by the method developed by Binnig and Rohrer [21] of IBM (Zurich) Labs in 1982. This method, which is exceedingly simple in principle, is based on monitoring the tunnel current between an atomically sharp tip and the surface as the tip scans the surface at a height of 10 Å or less.

Used in the constant current mode the tip can be shown to trace contours of the constant wavefunction overlap of the tip and the surface states contributing to tunnel current. These states can be selected by changing the sign and magnitude of the applied voltage between the tip and the surface. The STM thus provides an image of the local electronic structure of the surface apart from topography. Resolutions of 0.05 Å perpendicular to the surface and of 2-6 Å parallel to the surface can be obtained providing an extremely powerful surface diagnostic tool. Many fine examples of both electronic structure studies and adsorbate structure studies on surfaces have been provided. The strong impact of the discovery of STM on surface studies was recognized by the award of the 1986 Nobel Prize.

In addition, a number of other areas of semiconductor physics are being vigorously pursued which could not be included in these lectures due to time limitations. These include the upcoming field of semimagnetic superconductors currently attracting a lot of interest, 1-D and 0-D electron systems (quantum wires and dots), and polymeric semiconducting materials. All these developments bear testimony to the fact that semiconductor physics is a rich and highly fertile field holding tremendous promise for both new physics and new applications.

References

1. Tsu, R., Esaki, L., *Appl. Phys. Lett.*, vol. 22, p. 562, 1973.
2. Chang, L. L., Esaki, L., Tsu, R., *Appl. Phys. Lett.*, vol. 24, p. 593, 1974.
3. Dingle, R., Festkorperprobleme XV, *Advances in Solid State Physics*, 1975.
4. Narayanamurti, V., *Proc. 18th International Conf. on the Physics of Semiconductors*, vol. 1, Ed. by D. Angstrom, World Scientific, p. 3, 1987.
5. Shah, N. J., Pei, S. S., Tu, C. W., Tiberio, R. C., *IEEE Trans. Electron Devices*, ED-33, p. 543, 1986.

6. Capasso, F., Tsang, W. T., Williams, G., *IEEE Trans. Electron Devices*, ED-30, p. 381, 1983.
7. Temkin, H., Alavi, K., Wagner, W. R., Pearsall, T. P., Cho, A. Y., *Appl. Phys. Lett.*, vol. 42, p. 845, 1983.
8. Vincent, G., Bois, D., *Solid State Commun.*, vol. 27, p. 431, 1983.
9. Tkachev, V. D., Makarenko, L. F., Markovich, V. P., Murin, L. I., *Fiz. Tekh. Poluprovodn.*, vol. 18, p. 526, 1984.
10. Wruck, D., Spiegelberg, F., *Phys. Stat. Sol.*, (b) vol. 133, K39, 1986.
11. Wezep, D. A., Gregorkiewicz, T., Bekman, H. H. P. Th., Ammerlaan, C. A. J. *Proc. 14th International Conf. on Defects in Semiconductors*, Part 3, Ed. by H. J. von Bardeleben, Trans Tech Publications, p. 1009, p. 1986.
12. Lang, D. V., Logan, R. A., *Phys. Rev. Lett.*, vol. 39, p. 635, 1977.
13. Henning, J. C. M., Ansems, J. P. M., *Proc. International Conf. on Defects in Semiconductors*, Part 1, Ed. by H. J. von Bardeleben, Trans Tech Publications, p. 429, 1986.
14. Johnson, N. M., Herring, C., Chadi, D. J., *Phys. Rev. Lett.*, vol. 56, p. 769, 1986.
15. Pankove, J. I., Carlson, D. E., Berkeyheiser, J. E., Wance, R. O., *Phys. Rev. Lett.*, vol. 51, p. 2224, 1983.
16. Johnson, N. M., *Phys. Rev. B*, vol. 31, p. 5525, 1985.
17. Pantelides, S. T., *Proc. 14th International Conf on Defects in Semiconductors*, Part 2, Ed. by. H. J. von Bardeleben, Trans Tech Publications, p. 573, 1986.
18. Fritzsche, H., *Physics Today*, p. 1, Oct. 1984.
19. Döhler, G. H., *Proc. 17th International Conf. on the Physics of Semiconductors*, Ed. by J. D. Chadi and W. A. Harrison, Springer-Verlag, p. 491, 1985.
20. Kakalios, J., Fritzsche, H., *Proc 17th International Conf. on the Physics of Semiconductors*, Ed. by J. D. Chadi and W. A. Harrison, Springer-Verlag, p. 503, 1985.
21. Binnig, G., Rohrer, H., *Proc. 17th International Conf. on the Physics of Semiconductors*, Ed. by J. D. Chadi and W. A. Harrison, Springer-Verlag, p. 3, 1985.

SEMICONDUCTOR MATERIALS FOR PHOTOELECTROCHEMICAL CELLS AND PHOTOCATALYSTS

Kirthi Tennakone
Institute of Fundamental Studies, Hantana
Kandy, Sri Lanka
and
Department of Physics
University of Ruhuna
Matara, Sri Lanka

Abstract: **Photoelectrochemical cells and photocatalysts require semiconductor materials with special properties. However, as polycrystalline films and microcrystals can be prepared from a large variety of materials, it is possible that complex and unconventional materials find applications in these devices. Semiconducting, photochemical, photocatalytic properties of some novel materials and possible applications are discussed.**

Introduction

Photoelectrochemical cells (PEC) based on semiconductor electrodes are extensively investigated as possible systems for conversion of solar energy[1-7]. The advantage of PEC is that polycrystalline or amorphous materials can be used as the performance of these cells is less sensitive to defects and grain boundaries when compared to solid state cells. Again in addition to generation of electricity they have the capacity of producing energy rich compounds. However PECs are plagued with corrosion. Low band gap materials sensitive to the visible solar spectrum generally photodegrade and high band gap material that are photostable are insensitive to the visible spectrum. Some of the directions in which PECs could hopefully be stabilized are (1) search for efficient redox couples, that suppress corrosion reactions in preference redox reactions in the electrolyte (even the best redox couples are known to cause some corrosion in aqueous media, e.g., CdS in polycalcogenide solutions) (2) search for solid state structures that resist photocorrosion (e.g., layered structures) (3) sensitization of high band gap materials with dyes (4) use of non-aqueous electrolytes.

Another promising method for conversion of solar energy is the use of semiconductor photocatalysts for production of energy rich compounds. Electron-hole pairs created in semiconductor particles dispersed in an electrolytic medium by absorbed photons can participate in redox reactions fixing chemical energy. If conversion efficiencies are increased (this can be done if methods are found to initiate multi-photon processes) and simple methods are found to separate the oxidation reduction products particulate systems would turn out be one of the cheapest methods for storage of solar energy. The type of semiconductor materials adaptable is highly restrictive in the case of dry solid state cells. Semiconductor parameters such as band gap, optical absorbtion coefficient, carrier life times, mobilities and material properties like the ability to process samples of large surface area decide the suitability of a material for solid state cells. In wet cells a larger variety of polycrystalline and amorphous materials can be used. Photostability, a band gap appropriate to the solar spectrum, high carrier mobility and readily available methods for preparation of surface coatings are essential requirements of a material for photoelectrochemical use. In the case of a semiconductor photocatalyst, the above requirements can be further relaxed, there is no necessity for preparation of surface coatings, the carrier mobilities and recombinations become less important as the particle size decreases. Thus photostability and band gap are the major deciding factors of a photocatalyst (some investigations suggest that even the band gap may be an adjustable parameter for a semiconductor particle) As an example of an unconventional material with promising properties, the semiconducting and photoelectrochemical properties of β-CuCNS are described in detail.

Cuprous Thiocyanate

In a series of reports we have discussed the properties of CuCNS and its suitability as a material for dye-sensitized photoelectrochemical cells[8-11]. The solid CuCNS exists in two polymorphic forms, α-CuCNS with orthorhombic crystals and β-CuCNS that grow as trigonal pyramids[11-15]. The β-form is polytypic[14-15] and consists of 3R and 2H polytypes in syntatic coalescence[14-15]. The β-structure is composed of layers of closed-packed network of CuCNS units, the adjecent layers are bonded to each other by strong Cu-S bonds[14-15]. Cu and S are tetrahedrally coordinated and the crystal is a three dimensional polymer[15-16]. The α and β forms are readily identifiable by the CN stretching frequency in the infraed spectrum which occurs at 2157 cm^{-1} and 2173 cm^{-1} respectively[16].

CuCNS can be easily deposited on copper substrates by immersing them in solutions containing the CNS^- ion or by electrolysis. All such deposits are found to consist of the β-form. The polymeric nature of CuCNS with strongly interbonded layers has number of advantages. Films deposited copper surfaces are generally free from cracks and pin-holes[10], they are firmly affixed to the substrate and readily adsorbs other substances (notably those containing S or CNS ligand). We have found that copper doped β-CuCNS is n-type and sensitive to the visible spectrum, while the p-type which is insensitive to the visible spectrum originates from a stoicheiometric excess of CNS. The methods of preparation of n and p-type β-CuCNS, their photoelectrochemical and solid state properties are described below.

Method I (a) p-type β-CuCNS: Copper sheets (2 x 5 cm, Aldrich 99.99%) ultrasonically cleaned after anodic polishing in 10% H_2SO_4 were kept immersed in an aqueous solution 0.5 M in KCNS and 0.7 M acetic acid (pH ∿ 3.5). Propan-2-ol (5-10 ml per 100 ml of the solution) was added to keep the surface free from grease. Depending on the duration of immersion, films few microns to fraction of a millimeter were deposited (b) n-type β-CuCNS: Copper sheets cleaned as in (a) were kept immersed in an aqueous solution 0.05 M in KCNS, 0.3 M Na_2SO_3 acidified with acetic acid until pH ∿ 5 (propan-2-ol added as earlier). Deposition under these conditions was slow, a layer ∿ 10 microns formed in 6-8 h. To obtain a film ∿ 0.1 mm thick it was necessary to expose the plate to the solution 6-8 days, replenishing Na_2SO_3 frequently.

Method II (a) p-type β-CuCNS: An aqueous solution 0.25 M in KCNS, 0.7 M in acetic acid containing propan-2-ol was electrolysed with a copper plate as the anode. The current density was kept at ∿ 2.5 mA cm^{-2}. (b) n-type β-CuCNS: An aqueous solution 0.25 M in KCNS, 0.01 M in $CuSO_4$ (Containing propan-2-ol) was electrolysed with a copper plate as the cathode. The current density was kept at ∿ 1 mA cm^{-2}.

n-Type β-CuCNS films are chocolate brown in colour, those of p-type are white. Both the types are resistant to dilute mineral acids, but readily attacked by alkalis. The electrolytic method of deposition has several advantages. The growth is faster the thickness of the film is easily controlled and the level of doping can be adjusted by altering the concentration of CNS^- and Cu^+ ions.

The chemicals used in all experiments were of reagent quality. KCNS (Analar) was further purified by recrystallization.

The chemical composition of the deposits were determined by the following method: Plates were digested in a hot solution of NaOH (0.1 M) for 1 h to extract CNS in the form of NaCNS. The CNS^- ion concentration in the acidified extract was determined spectrophotometrically after addition of $FeCl_3$. The deposits were also dissolved in saturated KCNS solution (CuCNS is soluble in saturated KCNS) and the copper concentration in the extract was determined by the method described in ref. 17. The material carefully scraped from thick deposits using an agate razor was also analysed for K, Cu and other elements by atomic absorption spectroscopy (Hitachi 170-30). The IR spectra of the scraped powder was also obtained (Hitachi 270-50). Photoelectrochemical measurements were carried out under potentiostatic conditions (Hokuto Denko HA-301 Potentiostat) using a Pt counter electrode. Light intensities were measured with an International Light IL 700 Radiometer. Photocurrent spectra were ascertained with an Applied Photophysics Monochromator and 100 W tungsten filament lamp. Diffuse reflectance spectra were obtained with a Hitachi U-3200 Spectrophotometer. The electrolyte used was 0.01 M KCNS.

Single crystals and polycrystalline powder of β-CuCNS were also prepared by the methods described in refs. (15) and (12) respectively. The copper doped β-CuCNS powder was prepared by the following method

(Method III): Copper sponge (free from oxide) was kept suspended in a solution 0.05 M in KCNS, 0.3 M in Na_2SO_3 acidified with acetic acid (pH $\sim$ 5) for one week. The precipitate was filtered, washed with dil HNO_3 to remove axcess copper and dried at 110°C in a current of nitrogen. Attempts to dope CuCNS single crystals with copper was not successful.

The sign of majority charge carriers in single crystals and compressed powder pellets were determined by Hall tests. Electrical conductivity of the materials were measured using Takeda Riken Computing Digital Multimeter (TR 6877).

The IR spectra of p and n-type CuCNS have same structure with an intense band at 2173 cm^{-1} and a hardly distinguishable maxima in absorption at 2157 cm^{-1} (Fig. 1), clearly indicating that the polymorphic form depositing on copper is almost entirely the β-form. The chemical analysis and the sign of majority charge carriers in β-CuCNS prepared by different methods are presented in Table 1. It is clear that the n and p-type originate from stoicheiometric excess of Cu and CNS respectively. Strongly reducing environment,low concentration of CNS^- ions or electrolytic discharge of Cu^+ ions at the substrate favour formation of the n-type. High concentration of CNS^- ion or its rapid discharge at the copper substrate yield p-type films. The films of the p-type CuCNS change into the n-type upon several weeks of storage. This can be understood as resulting from diffusion of copper from the substrate into the film. The polycrystalline powder and the single crystals made by the methods described in the experimental section are p-type. The powder can be changed into the n-type by heating (150°C) with copper dust in a stream of N_2, we have not succeeded in doping single crystals (1-2 mm) of β-CuCNS with Cu or grow them under conditions that would incorporate a stoicheiometric excess of Cu.

The temperature variation of the d.c. electrical conductivity of p and n-type materials are indicated in Fig. 3. Distinct differences are seen, in the p-type CuCNS (excess CNS ∿ 2%) the conductivity initially increases and then begin to decrease at a relatively low temperature. The thermal activation energy of increase being 0.2 eV. However, in the n-type CuCNS (excess Cu ∿ 9%) rapid increase in conductivity with a higher activation energy ∿ 0.51 eV occurs at ∿ 143°C and continues until the decomposition temperature (340°C) is reached. The conclusion we can arrive at from the above observation is that the acceptor impurity levels in the p-type material are located close to the valence band (the decrease in conductivity occurs once they get fully populted), but in the n-type material the donor levels lie well below the conduction band (Fig. 3).

The diffuse reflectance spectra of the two types (Fig. 4) also supports the above conclusion. In the p-type material the sharp absorption adge seen at 345 nm corresponds to a band gap ≈ 3.6 eV. The n-type material also shows a similar absorption edge located approximately at the same position. In addition two other absorption edge like structures, an intense one at 630 nm (∿ 2.0 eV) and a relatively weak one at 980 nm (∿ 1.3 eV) are also noticeable. Undoubtedly the more intense edge at 630 nm arise as a result of the transitions to the conduction band from the donor impurity (Cu) band. The weaker band originates as a result of the transitions from valence band to the Cu impurity band.

The photocurrent spectra of p and n-CuCNS in aqueous KCNS are given in Fig. 5. In the case of p-CuCNS the photoresponse is insignificantly small in the visible region, but shows a sharp peak near 340 nm. On the other hand n-CuCNS responds to visible spectrum and show three peaks at 340 nm, 648 nm and 980 nm corresponding to band gap excitation, transitions from impurity band to the conduction band and transitions from valence band to the impurity band respectively. The absoption corresponding to the band gap excitation is weak because, electrons excited from the valence band are also trapped in the impurity band. The most intense peak results from impurity band to conduction band transitions.

The photocurrent potential curves for the two types of CuCNS electrodes in $\sim 10^{-2}$ M KCNS (pH $\sim$ 7) are displayed in Fig. 6. In the case of the n-type material, the on-set occurs at - 0.78 V vs SCE, whereas in the p-type, the photocurrent becomes zero at - 0.64 V vs SCE and the response become cathodic upon further decrease of the biasing potential.

CuCNS is chemically stable in acidic media, it also shows a remarkable photostability in aqueous KCNS (Aqueous KCNS acts as a recox couple owing to the existence of CNS^- as well as $(CNS)_2^-$ ions (8)) . Fig. 7 shows the development of the photocurrent in n-type CuCNS photoanode baised - 0.03 V vs SCE. The I-V characteristics of the cell are shown in Fig. (8), the energy conversion efficiency is quite small (4 x 10^{-3}, light source 100 W tungsten filament lamp).

The unusual structural features, photostability and the possibility of sensitization to visible spectrum by doping makes ≠CuCNS and interesting material which can even find applications in practical devices. The preparation of single crystal films and controlled doping could improve the conversion efficiencies. Unfortunately the heavy doping required for sensitization could also have undesirable effects, the impurities could act at recombination centres.

CuCNS readily adsorb thiocyanates of cationic dyes and consequently an ideal material for studying dye sensitization[9,11,18]. Sensitized catalysts capable of sacrificial photo-oxidation and reduction water can be prepared from this material[19-20]. The related AgCNS was also found to have semiconducting and photocatalytic properties. However this is material photounstable and readily decompose into metallic silver.

References

1. Gerischer, H., *J. Electronalytical Chem.*, vol. 58, no. 263, 1975.
2. Morrison, S. R., *Electrochemistry at Semiconductor and Oxidized Metal Electrodes*, New York: Plenum Press, 1981.
3. Heller, A. Ed., *Semiconductor Liquid Junction Solar Cells*, New Jersey: The Electrochemical Society, Princeton, 1977.
4. Fujishima, A., Honda, K., *Bull. Chem. Soc. Jpn.*, vol. 44, p. 1148, 1971.
5. Geisecher, H., In: *Solar Energy Conversion, Topics in Applied Physics*, vol. 31, Ed. by B. O. Seraphin, Berlin: Springer-Verlag, 1979.
6. Wallace, W. L., Nozik, A. J., Deb, S. K., Wilson, R. H. Eds., *Photoelectrochemistry Fundamental Processes and Measurement Techniques*, New Jersey: The Electrochemical Society, Princeton, 1982.
7. Tributsch, H., *Ber Bunsenges, Phys. Chem.*, vol. 81, p. 361, 1977.
8. Tennakone, K., *J. Phys. D*, vol. 16, p. L5, 1985.
9. Tennakone, K., Kahanda, M., Kasige, C., Abeysooriya, P., Wijenayake, R. H., Kaviratne, P., *J. Electrochem. Soc.*, vol. 131, p. 1575, 1984.
10. Tennakone, K., Shantha Pushpa, G. S., Punchihewa, S., Epa, G., *Electrochimica Acta*, vol. 31, p. 315, 1985.
11. Tennakone, K., Karunamuni, J., Dewasurendra, M., *Solar Energy Materials*, vol. 13, 447, 1986.
12. Kruger, D., Tschirch, E., *Chem. Ber B*, vol. 74, p. 1378, 1941.
13. Garaj, J., Gazo, J., *Chem. Zuesti*, vol. 19, p. 593, 1956.
14. Kabesova, M., Dunaj-Jurco, M., Serator, M., Gazo, J., Garaj, J., *Inorg. Chim,. Acta*, vol. 17, p. 161, 1976.
15. Smith, D. L., Saunders, V. I., *Acta Cryst. B*, vol. 37, p. 1807, 1981.
16. Smith, D. L., Saunders, V. I., *Acta Cryst. B*, vol. 38, p. 907, 1982.
17. Vogel, A. I., *Text Book of Quantitative Inorganic Analysis* 4th Rev., Ed. by J. Bassett, R. C. Denney, G. H. Jeffery, J. Mendham, 1985.
18. Tennakone, K., Dewasurendra, M., Kariapper, M. S., *Solar Energy Materials*, vol. 14, p. 499, 1986.
19. Tennakone, K., Shantha Pushpa, G. S., Basu, S., *Chem. Phys. Lett.*, vol. 121, p. 155, 1985.
20. Tennakone, K., Wickramanayake, S., Gunasekara, M. U., *J. Chem. Soc. Faraday Trans.* 1, vol. 83.

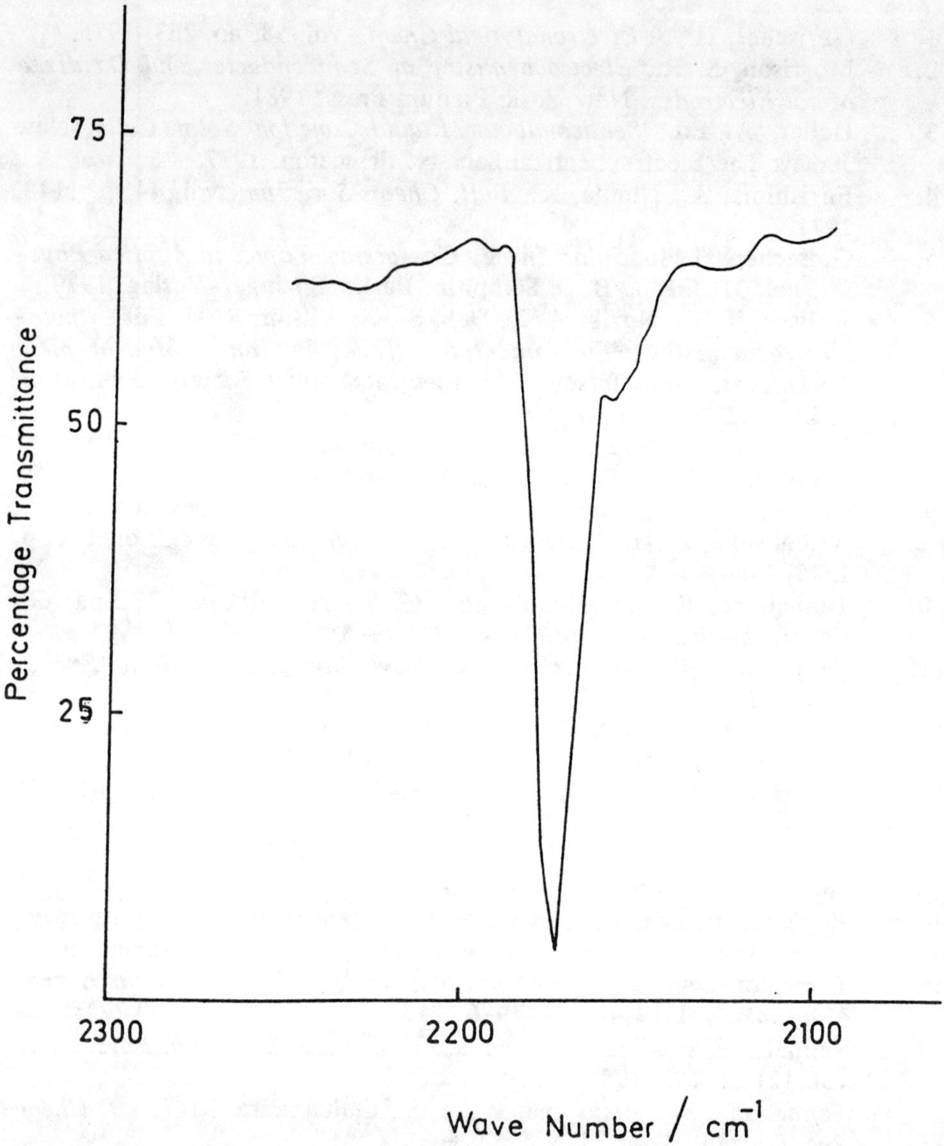

Fig. 1. IR spectra of CuCNS deposited on copper plates. Absorption maxima at 2173 cm^{-1} and 2157 cm^{-1} corresponds to β and α forms.

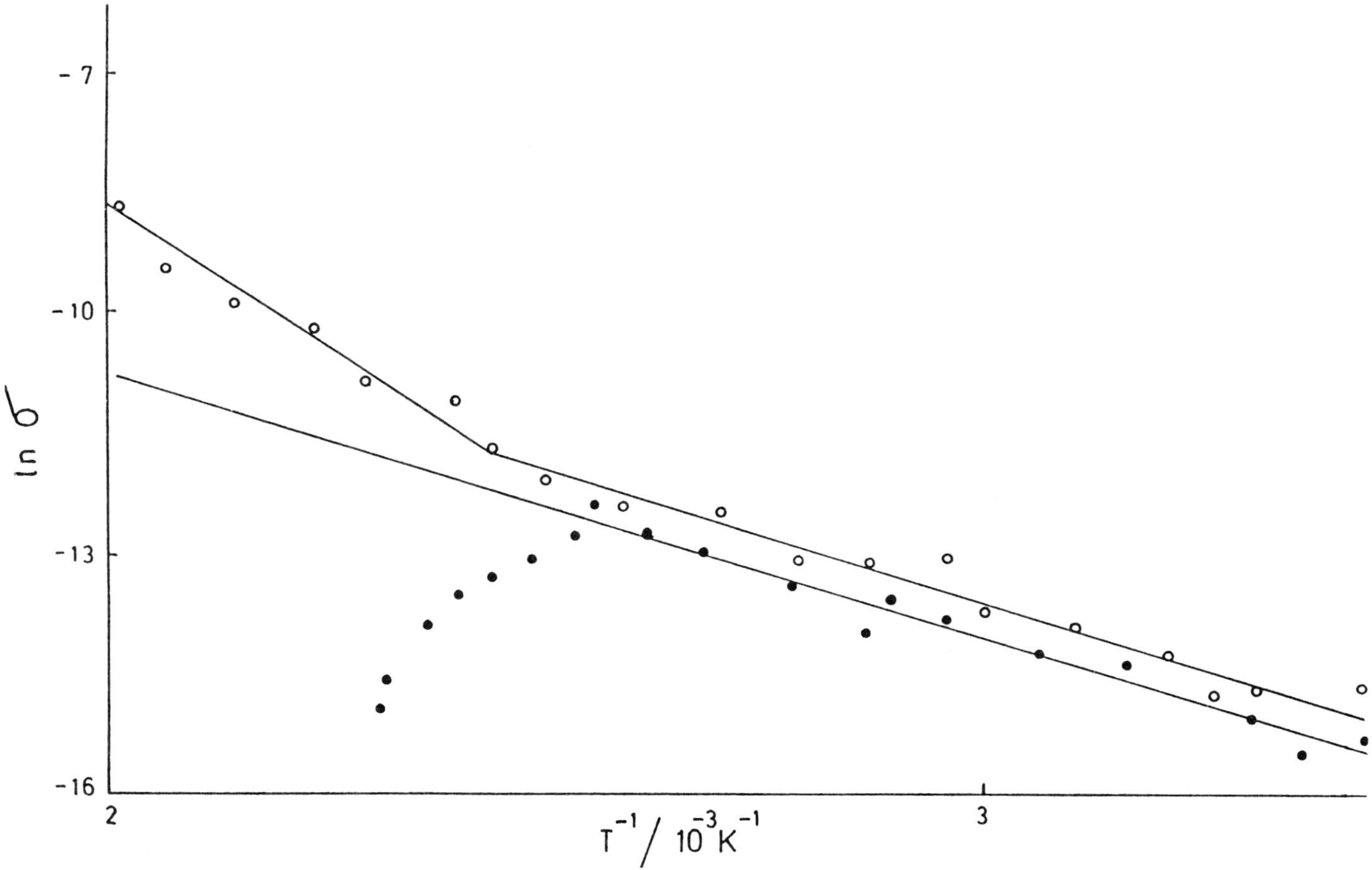

Fig. 2. Plot of lnσ (σ in Ω^{-1} m^{-1} vs T^{-1} for β-CuCNS • - p-type, ○ - n-type.

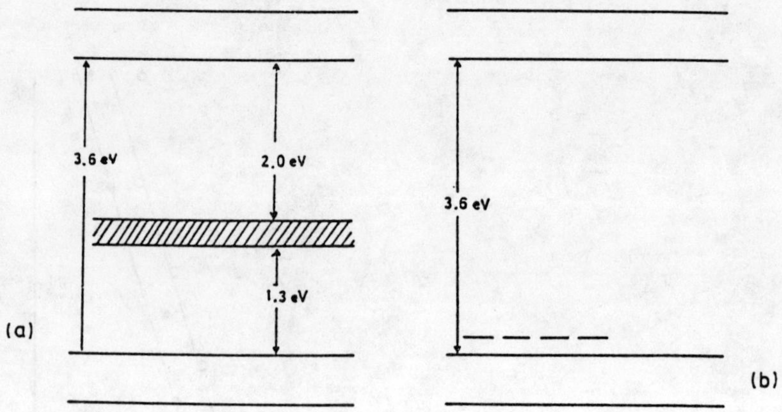

Fig. 3. Energy bands and impurity levels in (a) n-type and (b) p-type β-CuCNS.

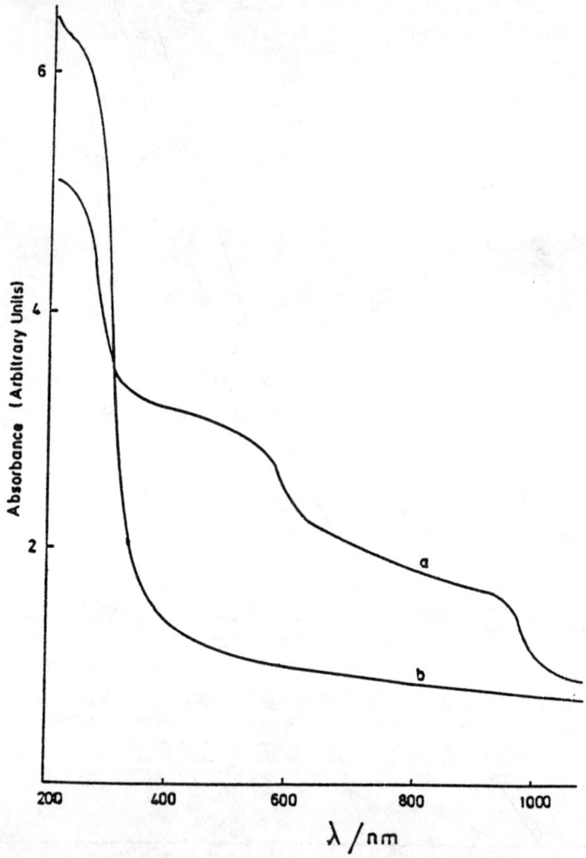

Fig. 4. Diffuse reflectance spectra of (a) n-type and (b) p-type β-CuCNS

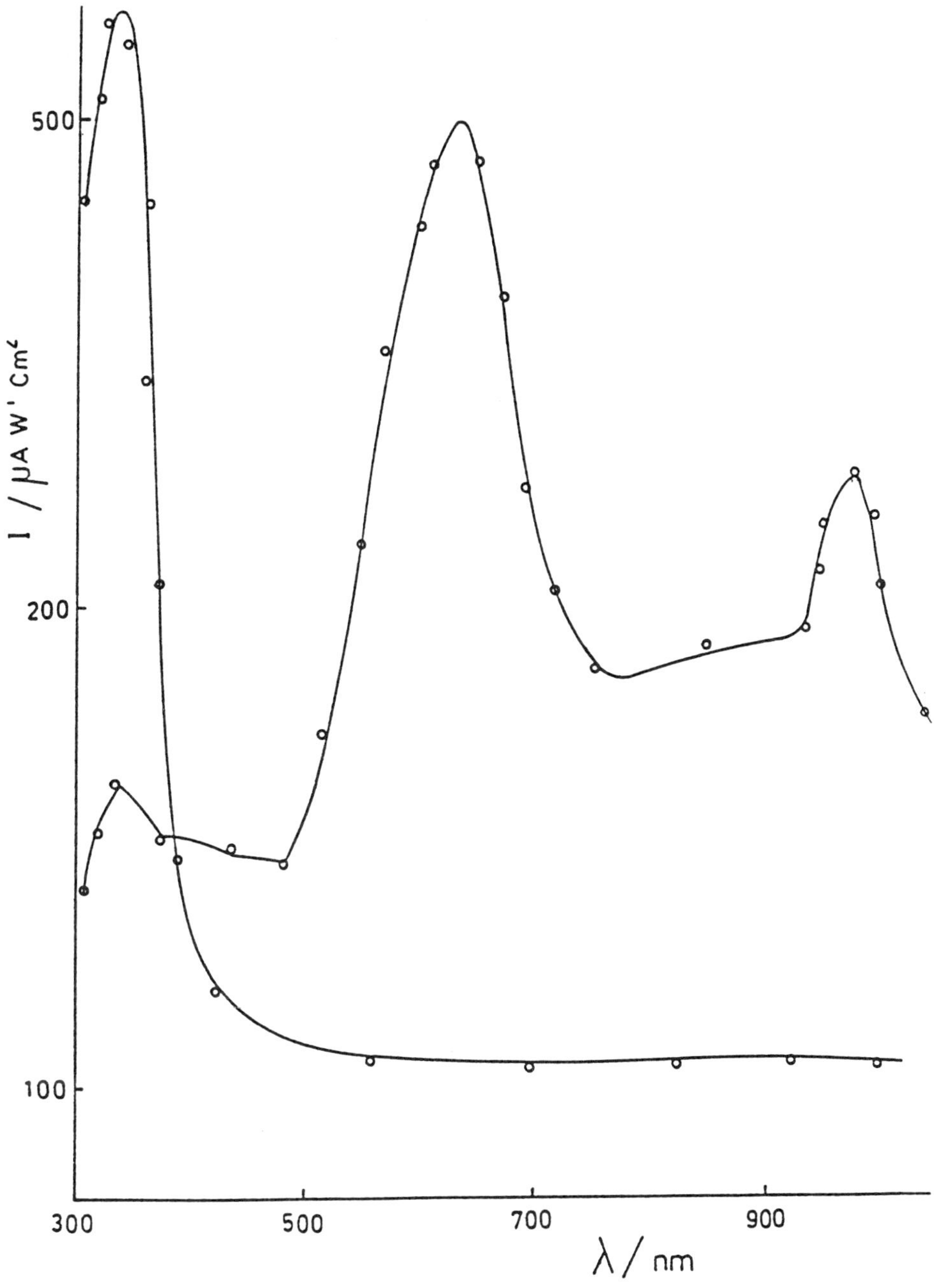

Fig. 5. Photocurrent spectra of (a) n-type and (b) p-type β-CuCNS.

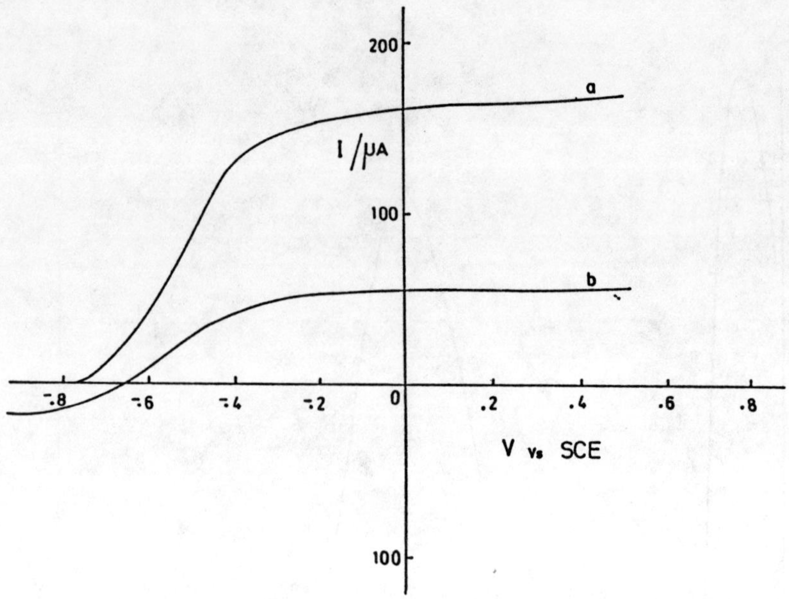

Fig. 6. Photocurrent-potential curves for (a) n-type and (b) p-type β-CuCNS (visible light).

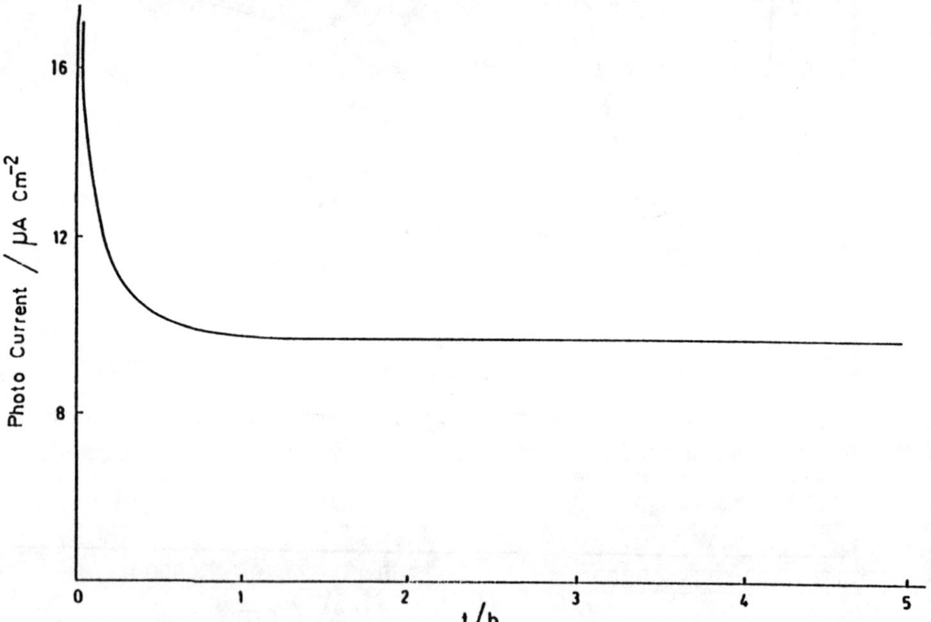

Fig. 7. Time development of the photocurrent in an n-type α-CuCNA photocathode biased -0.03 V vs SCE.

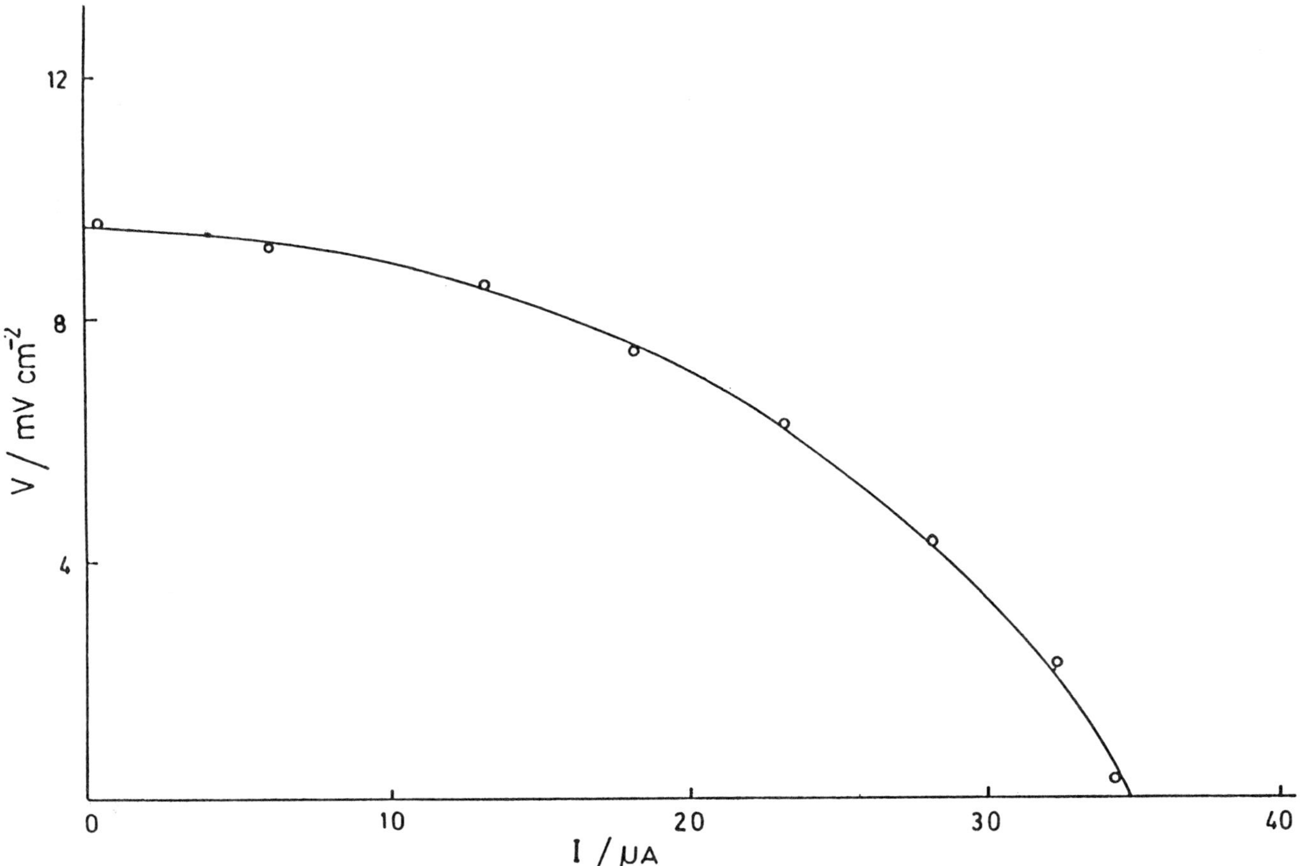

Fig. 8. V–I characteristic of the n-CuCNs/0.1 M KCNS/Pt cell (light source 60 W tungsten filament lamp. Biasing voltage -0.03 V vs SCE)

Table 1
Chemical analysis and sign of the majority charge carries in α-CuCNS prepared by different methods.

Preparation	Analysis	Type
Method I (a)	Cu(47.6%), CNS(51.8%), K(< 0.5%)	p
Method II (a)	Cu(48.0%), CNS(51.2%), K(~ 0.6%)	p
Polycrystalline Powder	Cu(48.2%), CNS(50.9%), K(~ 0.7%)	p
Method I (b)	Cu(57.1%), CNS(42.2%), K(~ 0.6%)	n
Method II (b)	Cu(56.0%), CNS(42.8%), K(~ 0.1%)	n
Method III	Cu(58.8%), CNS(40.7%), K(< 0.5%)	n

STRUCTURAL AND SYNTHETIC ASPECT OF SOLIDS

J. Gopalakrishnan
Solid State and Structural Chemistry Unit
Indian Institute of Science
Bangalore-560012, India

Abstract: **Chemists have contributed significantly many recent developments in solid state science (high-temperature superconductors, superionic conductors, novel heterogeneous catalysts, advanced ceramics of ferroelectrics) by synthesizing new solid and developing novel methods for solid synthesis. They have also tried to provide new ways of describing the structures of solids to show the interrelationships between crystal structure and chemical bonding. These lectures will present some of the recent developments in crystallography and chemical bonding. In the first lecture, the structures of metal and alloys will be discussed in the light of the Engle-Brewer correlation, Miedema's theory, and Chelikoswky and Phillips theory. Limitations of the conventional approach towards the structures of nonmolecular (ionic) solids based on ionic bonding and ionic radii will be pointed out. The use of orbital radii of atoms, as derived from pseudobinary solids will be discussed.**

In the second lecture aspects of designing new solids with specific structures and properties based on the knowledge of crystallography and reactivity will be examined. Among the several strategies for the synthesis of new solids, methods based on topochemical reactions and precursor compounds are noteworthy. A number of examples of recent synthesis of novel solids, mainly oxides and chalcogenides of transition metals, will be discussed. The design of new layered oxides exhibiting ion-exchange and intercalation behavior will also be presented.

STRUCTURAL CHEMISTRY OF POROUS FRAMEWORK SILICATES

R.P. Gunawardana
Department of Chemistry, University of Peradeniya
Peradeniya, Sri Lanka

Introduction:

Silicates are of tremendous importance to solid state scientists because most of the naturally occurring minerals (nearly 80%) are silicates of one form or another. They comprise both naturally occurring and synthetic or manufactured substances and include some of the earliest materials used by man, eg. pottery, clay, cements, cremamics and glass.

The fundamental characteristic of all silicate structures is the tetrahedral coordination of silicon by oxygen. As such, TO_4 tetrahedren (T=Si, Al etc) can be regarded as the basic structural unit in all silicates.

In general silicates are classified according to various ways in which TO_4 tetrahedra are linked to form polymeric anions. Thus, the silicates are classified into the following classes.

1. Orthosilicates containing isolated TO_4 groups eg. olivines, garnets.
2. Pyrosilicates eg. $Ca_2MgSi_2O_7$.
3. Chain silicates eg. pyroxenes.
4. Ring silicates eg. beryl, cordierite.
5. Layer or sheet silicates (2D) eg. kaolin, mica.
6. Framework silicates (3D) or tectosilicates eg. quartz, zeolites.

Framework Silicates (Tectosilicates):

Framework silicates have 3-dimensional 4-connected networks of SiO_4 tetrahedra in which all four oxygens are shared between adjascent tetrahedra.

These tectosilicates can be classified by its framework density[1] (or tetrahedral density), d_f, which is usually measured as number of (TO_4) tetrahedra, N_t, per 1000 $A°^3$

$$d_f = \frac{N_t \times 1000}{V_{cell}}$$

where V_{cell} = volume of unit cell in $A°^3$, N_t = number of tetrahedral atoms (Si, Al) in the unit cell. The framework density varies from 29.3 silicon atoms in the high pressure silica polymorph coesite to about 12.7 (Al + Si) atoms per 1000 $A°^3$ in zeolite, Faujasite. Depending on the magnitude of d_f, tectosilicates can be divided[2] into two classes namely, dense tectosilicates and porous tectosilicates.

Dense Tectosilicates:

Framework silicates with densities $d_f \gtrsim 21$ n_t per 1000 $A°_3$ have only narrow interstices between neighbouring atoms. These interstices are either empty (eg. quartz, cristobalite etc) or contain manatomic cations and sometimes water or other small neutral molecules (eg. feldspar, nepheline). These are referred to as dense tectosilicates.

Porus Tectosilicates:

Tectosilicates with tetrahedral densities $d_f < 21$ n_t/1000 $A°^3$ have larger voids between the TO_4 tetrahedra. These voids are usually occupied by neutral molecules or ions. Irrespective of the shape of these voids, these tectosilicates are referred to as porous tectosilicates. Porous tectosilicates may be subdivided into two classes depending on the nature of voids namely, (i) Clathrasils and (ii) Zeosils (isostructural with zeolites). A list of porous tectosilicates and their d_f values are given in Table 1.

(i) Clathrasils[3]: If the voids have the shape of concave polyhedra with "windows" too small to let the void-filling species (guest molecules) pass, then the guest species are enclathrated within these polyhedra as in a cage. The guest molecules cannot leave the cages, unless they are decomposed into smaller fragments or the cages are disrupted. Such tectosilicates containing only SiO_4 tetrahedra and the guest species, are referred to as clathrasils.

Table 1	
Framework densitites (d_f) of different tectosilicates	
Phase	d_f
Dense tectosilicates	
Coesite	29.3
Quartz	26.6
Keatite	25.1
Cristobalite	23.2
Tridymite	22.9
Porous tectosilicates	
(i) Clathrasils	
Melanophlogite	19.0
Dodecasil 1H	18.5
Dodecasil 3C	18.6
Deca-dodecasil 3R	17.6
Nonasil	19.3
Silica-sodalite	17.4
(ii) Zeosils	
Silica-ferrierite	19.3
Silica-ZSM-22	19.7
Silica-ZSM-48	19.9
Silica-ZSM-5	17.8
Silica-ZSN-11	17.9

Table 2 : Data on different clathrasils

Clathrasil family	Crystal system	Space group, cell dimensions (A°)	Cage type, volume and number of cages per unit cell		
Melanophlogites	Cubic	Pm3n a = 13.783	2 $[5^{12}]$ ($V \approx 80$ $A°^3$)		6 $[5^{12}6^{2}]$ ($V \approx 160$ $A°^3$)
Dodecasils 3C	Cubic	Fd3 a = 19.402	16 $[5^{12}]$		6 $[5^{12}6^{4}]$ ($V \approx 250$ $A°^3$)
Dodecasils 1H	Hexagonal	P 6/mmm a = 13.783 b = 11.190	3 $[5^{12}]$	2 $[4^{3}5^{6}6^{3}]$ ($V \approx 50$ $A°^3$)	1 $[5^{12}6^{8}]$ ($V \approx 430$ $A°^3$)
Deca-dodecasil 3R	Rhombohedral	R $\bar{3}$m a = 13.860 c = 40.891	6 $[4^{3}5^{6}6^{1}]$ ($V \approx 35$ $A°^3$)	9 $[5^{12}]$	6 $[4^{3}5^{12}6^{1}8^{3}]$ ($V \approx 350$ $A°^3$)
Nonasils	Orthorhombic	Fmmm a = 22.323 b = 15.058 c = 13.627	8 $[5^{4}6^{4}]$ ($V \approx 25$ $A°^3$)	8 $[4^{1}5^{8}]$ ($V \approx 30$ $A°^3$)	4 $[5^{8}6^{12}]$ ($V \approx 290$ $A°^3$)
Silica-sodalites	Cubic	a = 8.836	2 $[4^{6}6^{8}]$		

(ii) Zeosils: In contrast, the tetrahedral framework may contain systems of polyhedral cavities interconnected by "windows" large enough for the guest species to pass through. When the windows are sufficiently wide, the interconnected cavities form channels (figure 2) through which the guest species can move freely. These silicates are referred to as zeolites when there is some aluminium substitution in the framework and the charge balance is achieved by the accommodation of cations in the channels. On the other hand, the silica end members of zeolites, which contain electrically neutral frameworks and neutral guest molecules are referred to as zeosils[2].

Structural Features of Clathrasils:

Interest for silica clathrates (clathrasils) started after the chemical investigations of the rare mineral melanophlogite. The structure of this mineral was determined by Gies and Liebau[5] (1983) who found the presence of 3-dimensional four connected network of SiO_4 tetrahedra with two types of cages : $2[5^{12}]$ and $6\ [5^{12}6^{2}]$ cages per unit cell. From difference fourier maps $[\rho_{obs}-\rho_{calc}(SiO_2)]$, it was concluded that the voids are occupied by the guest molecules such as N_2, CO_2, CH_4 etc.

Subsequently, five more clathrasil types have been reportd. They are silica-sodalite[6], dodecasil 3C[7], dodecasil 1H[8], deca-dodecasil 3R[9] and nonasil[10]. Framework types of these clathrasils along with their unit cell data and approximate cage volumes are given in Table 2.

All these clathrasils have been synthesized from homogeneous silicic acid solution prepared by hydrolyzing TMOS, $Si(OCH_3)_4$ in aqueous ethylenediamine solution. These solutions were filled into silica tubes and after addition of the guest species the tubes were sealed and subjected to thermal treatment at 160-240°C. The guest molecules which give the different clathrasils and the characteristic morphologies shown by these clathrasils are given in Table 3.

Since the guest molecules act as templates during synthesis, it is evident that the size and shape of the guest molecules control the size and shape of the cages in clathrasils. In addition, the following factors also play an important role in the crystallization of different clathrasils.

Table 3

Guest species for clathrasil synthesis

Clathrasil	Characteristic Morphology	Guest Species
Melanophlogite	Cubic	N_2, CO_2, CH_4, Kr, Xe; CH_3NH_3
Dodecasil 3C	Octahedral	CH_3NH_2, $C_2H_5NH_2$, $(CH_3)_2NH$ $(CH_3)_3N$, $(C_2H_5)_2NH$, Pyrrolidine, piperidine
Dodecasil 1H	Hexagonal	Piperdine, 4-methylpiperidin 1-azabicyclooctane 1-aminoadamantane (high temperature)
Deca-dodecasil 3R	Rhombohedial	1-aminoadamantane (low temperature)
Nonasil	Lath-like	1-aminobutans 2-aminopentane

Figure 1

Changes in clathrasils

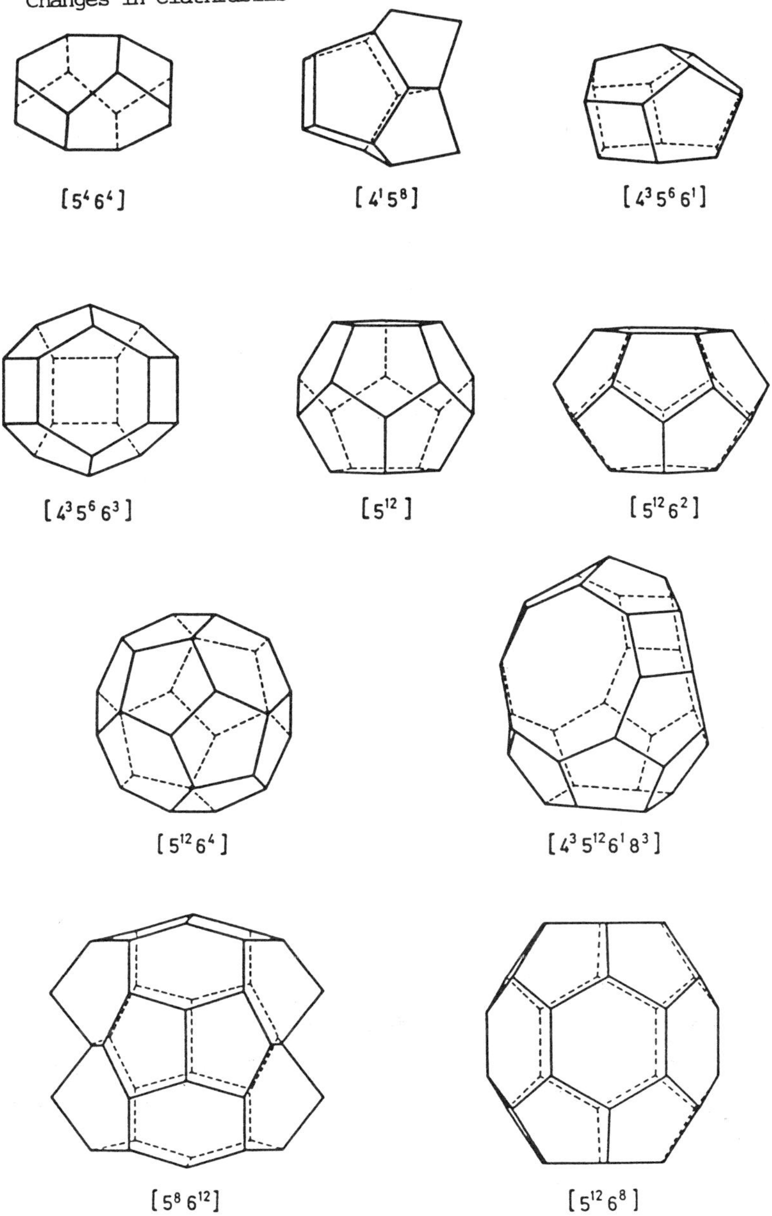

(a) Stability of the guest molecules

(b) Rigidity of the guest molecules

(c) Symmetry of the guest molecules

(d) Basicity of the guest molecules

(e) Temperature and pressure of synthesis

The type of cage present in different clathrasils are shown in figure 1. All known clathrasil frameworks contain two or more types of cages (Table 2). Larger guest molecules which occupy larger cages are considered as the structure controlling guests[11]. In contrast, the smaller cages may either be empty or be filled with small molecules. For instance, in melanophlogite larger $[5^{12}6^{2}]$ cages are occupied by CH_3NH_2, CO_2, N_2O or N_2. The smaller $[5^{12}]$ cages in contrast, are occupied by CH_4, N_2, Kr or Xe or they may remain empty. Formation of dodecasil 1H requires the presence of relatively large molecules such as piperidine, 1-aminoadamantane etc. which fill the larger $[5^{12}6^{8}]$ cages whereas smaller $[5^{12}]$ and $[4^{3}5^{6}6^{3}]$ cages are occupied by Nitrogen or they may be empty depending on the presence or absence of air during synthesis. Similarly, the larger $[5^{8}6^{12}]$ cages of the nonasils are filled with 2-aminopentane etc. (Table 3) whereas their $[5^{4}6^{4}]$ and $[4^{1}5^{8}]$ cages are empty because they are too small to accommodate even small gaseous molecules.

The nature of interaction between the structure controlling guest and the host framework is mainly of van der Waals type. In addition, it is also believed that the nitrogen atoms of the structure controlling guest molecules interact with silicon atoms of the framework by the formation of dative-type bonding.

Structural Features of Zeosils:

A series of high silica zeolites has been synthesized in the recent past. They include ZSM-5, ZSM-11, ZSM-48, ZSM-22 etc. A few silica and members of zeolites referred to as zeosils have also been synthesized recently. High silica zeolites are thermally very stable and they are of tremendous importance in industrial catalysis. Zeosils too are of considerable importance in understanding the properties of zeolites and in the interpretation of their NMR data. They are hydrophorbic, organophilic and may also be used as catalyst carriers.

Channel systems in zeolites

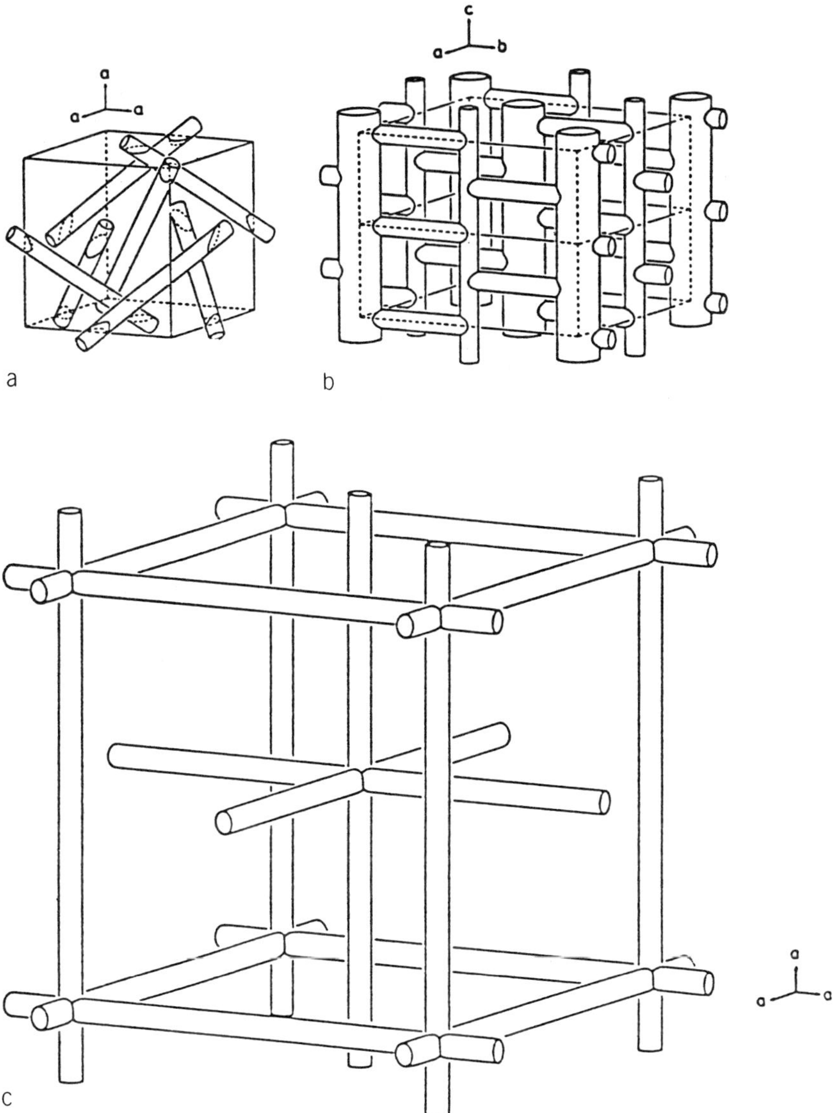

Figure 2

a. System of nonintersecting one-dimensional channels in analcime. $Na[AlSi_2O_6]\bullet H_2o$, running parallel to the <lll> zone axes (Ferraris et al. 1972). **b.** System of two-dimensionally intersecting channels parallel to (100) in mordenite. $Na[AlSi_5O_{12}]\bullet 3H_2O$ (Meier 1961). **c.** Two interpenetrating systems of three-dimensionally intersecting channels in paulingite. $(K.Ca.Na)_2[Al_3Si_{11}O_{28}\}\bullet 12H_2O$ (Gordon et al. 1966

Zeosils have been synthesized from aqueous silicic acid solutions using long chain polyamines and in some cases amine-boric acid pairs as templates. Long chain amines such as 1-aminobutane, diethylamine and 1,3 diaminopropane give silica-ZSM-22 with narrower 1D channel systems while diethylenetriamine, triethylenetetramine and tetraethylenepentamine act as templates for the formation of silica-ZSM-48 with wider 1D channel systems. On the other hand, highly branched amines such as triethylamine, tripropylamine and tributylamine form silica-ZSM-5 (silicalite-1) with 3D channel systems. Ethylenediamine in the presence of boric acid give rise to a zeosil possessing 2D channel systems namely, silica-fernerite[13]. In this case it is believed that theamine-boric acid acid-base pair acts as the template. All these zeosils are isostructural with their corresponding zeolites. Different types of channel systems present in zeolites are shown in figure 2. Zeosil framework is neutral unlike in the zeolites and as such only neutral species can act as templates for their synthesis.

On thermal treatment of clathrasils as well as zeosils, the organic species (templates) are broken down, oxidised and removed from the cages or channels without damaging the framework. Thus, these silica-frameworks have remarkable metastability upto about 1000°C even after the removal of the templates.

Although a number of clathrasils and zeosils have been synthesized and their thermal properties determined recently, their properties of industrial importance have not been investigated yet. It hoped that the solid state scientists would find useful applications of these porous materials in the near future.

References

1. Liebau, F., *Structural Chemistry Silicates*, Heidelberg: Springer-Verlag, 1985.
2. Liebau, F., Gies, H., Gunawardane, R. P., Marler, B., *Zeolites*, vol. 6, p. 373, 1986.
3. Gies, H., Liebau, F., Gerke, H., *Angew. Chem.*, vol. 94, p. 124, 1982.
4. Gies, H., Liebau, F., *Acta Crystallogr. A*, vol. 37, p. 187-188, 1981.
5. Gies, H., *Z. Kristallogr.*, vol. 164, p. 247, 1983.
6. Bibby, D. M., Dale, M. P., *Nature*, vol. 317, p. 157, 1985.
7. Gies, H., *Z. Kristallogr.*, vol. 166, p. 11, 1984.
8. Gerke, H., Gies, H., *Z. Kristallogr.*, vol. 166, 11, 1984.
9. Gies, H., *Z. Kristallogr.*, in press.
10. Maler, B., Dehnbostel, N., Eulest, H. H., Gies, H., Liebau, F., *J. Inclusion Phenomena*, in press.
11. R. P. Gunawardane, Gies, H., Liebau, F., *Z. Anorg. Alleg. Chem.*, in press.
12. Gunawardane, R. P., Gies, H., Maler, B., *Zeolites*, in press.
13. Gies, H., Gunawardane, R. P., *Zeolites*, in press.

References

SULPHATE BASED SOLID ELECTROLYTES: PROPERTIES AND SOME APPLICATIONS

M. A. Careem
Department of Physics
University of Peradeniya
Peradeniya, Sri Lanka

Abstract: **Pure Li_2SO_4 undergoes a phase transition at 577°C to become a highly conducting solid electrolyte with ionic conductivity of the order of 1 X^{-1} cm^{-1}. By mixing sulphates of different types in the right proportions highly conducting phases can be obtained at relatively lower temperatures. Characteristic of sulphate based systems is that both mono- and divalent cations are very mobile, whereas in other kinds of solid electrolytes the high mobility is restricted to only one type of cation. The high conductivity in f. cc. Li_2SO_4, bcc $LiNaSO_4$, bcc $LiAgSO_4$, and noncubic $Li_4Zn(SO_4)_3$ are caused by an unusual transport mechanism called the "paddle wheel mechanism" where the mobility of cations is strongly enhanced by a rotational motion of the transitionally static sulphate ions. Mechanisms associated with point defects dominate in phases like hexagonal Na_2SO_4 and monoclinic Li_2SO_4. Characteristic for them is that the conductivity can be increased several orders of magnitude by doping with divalent cations a in the case of $Li_{0.22}Na_{1.33}Zn_{0.22}SO_4$ which has a conductivity of 0.25 X^{-1} cm^{-1} at 560°C. The sulphate based solid electrolytes can be used in high energy density primary cells and in a heat storage system.**

SULPHATE BASED SOLID ELECTROLYTES: PROPERTIES AND SOME APPLICATIONS

Department of Physics

SOME PROBLEMS CONCERNING DISPERSING SYSTEMS

Chen Liquan
Institute of Physics, Academia Sinica
Beijing, China

Abstract: **Phase transition and phase relation in some dispersion ionic conductor systems are reviewed. The importance of multiphase composition in searching for new high temperature superconductors is emphasized.**

1. Introduction

A two-phase or a multiphase system is called a dispersion system. It is usually obtained by dispersing a matrix phase into fine second phase particles. The properties of a dispersion system are normally different from those of the individual phases. For instance, the ionic conductivity of the dispersion system LiI-Al_2O_3 is at least two orders of magnitude higher than that of pure LiI. We have already confirmed the ionic conductivity enhancement in many kinds of dispersion conductors [1].

More interesting are phase transitions and phase relations in a dispersion system which are discussed in this review.

2. Phase Transition of AgI and α-Fe_2O_3 in an AgI (α-Fe_2O_3 or η-Al_2O_3) Dispersion System

AgI is a typical ionic conductor. It changes from β- to α-phase at 147°C and vice versa. We have performed both electrical conductivity measurements and Digital-to-analog on AgI (η-Al_2O_3 or α-Fe_2O_3) [2]. A general feature of this dispersion system is the occurrence of hysteresis in log 6 vs. 1/T plot. The width of the hysteresis loop observed for AgI (50 m/Oη-Al_2O_3) corresponds to 17°C. Phase transition temperature from β- to α-phase is not evidently influenced, while the phase transition temperature from α- to β-phase is significantly lowered by the existence of η-Al_2O_3 or α-Fe_2O_3. This result was confirmed by DTA (Figure 1).

α-Fe_2O_3 has a Morin transition at 260 K. We therefore wonder whether or not the Morin transition of Fe_2O_3 is also affected by the matrix phase AgI. Figure 2 shows the variation of the Morin transition temperature vs. AgI content. It can be seen that the Morin transition temperature of α-Fe_2O_3 (AgI) is much lower than that of pure α-Fe_2O_3.

From the above results it seems clear that the phase transition temperatures of the individual phases in the dispersion system AgI (α-Fe_2O_3) has been influenced by the presence of the other phase.

3. Phase Transition of $LiIO_3$ in the $LiIO_3$ (γ-Al_2O_3) Dispersion System

The phase transition of $LiIO_3$ is rather complicated. A systematic study of this problem was made by my colleagues [3-10].

The variation of the phase transition temperature of $LiIO_3$ (γ-Al_2O_3) with γ-Al_2O_3 content is shown in Figure 3. At the beginning of the phase transition the temperatures of both the α- to γ- and γ- to β-phases have

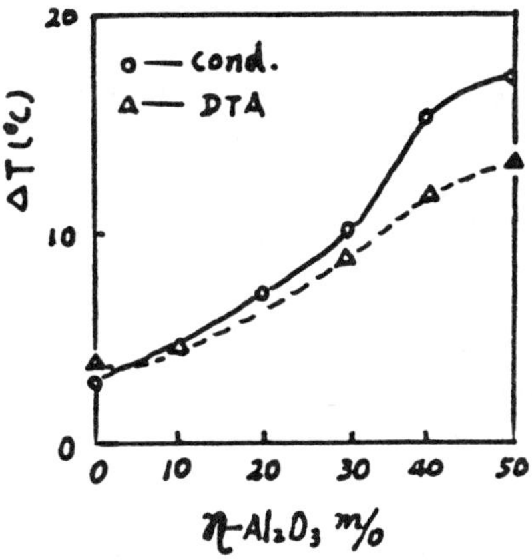

Fig. 1. Delay of α- to β-phase transition temperature vs. η-Al_2O_3 content.

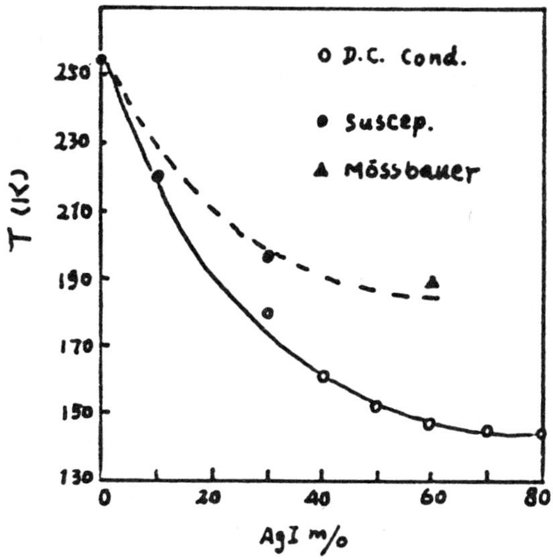

Fig. 2. The Morin transition temperature vs. AgI content.

Fig. 3. Phase transition temperature of $LiIO_3$ (γ-Al_2O_3) as a function of γ-Al_2O_3 content.

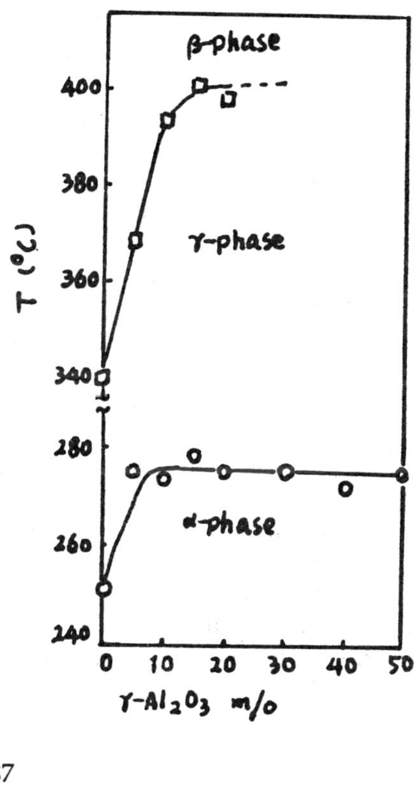

been increased remarkably by the addition of Al_2O_3. The transition temperature from the α- to the γ-phase has been increased from 250°C for pure $LiIO_3$ to 273°C for $LiIO_3$ (10 m/O γ-Al_2O_3) (an increment of 23°C). The transition temperature from γ- to β-phase has been increased from 340°C for pure $LiIO_3$ to 400°C for $LiIO_3$ (15 m/O γ-Al_2O_3) (an increment of 60°C). When the γ-Al_2O_3 content is more than about 10 m/O the phase transition temperature in both cases does not show a further increase.

Moreover the phase relation of $LiIO_3$ has been changed by the existence of γ-Al_2O_3. On heating pure α-$LiIO_3$ changes gradually to a γ-phase, β-phase and then melts. On cooling it changes back to an α-phase by passing through δ- and θ-phases after solidification. The β-phase could be obtained by cooling the sample as soon as the endothermic peak appears on Digital-to-analog heating process. A metastable ξ-phase can be obtained by special thermal treatment, i.e., by first thoroughly melting the $LiIO_3$ sample, then cooling to 200-220°C and heating again to about 385°C and finally cooling to room temperature. By quenching at high pressure the ξ-phase is easily obtained. Therefore, the ξ-phase is a high pressure phase.

In the $LiIO_3$-Al_2O_3 dispersion system when the Al_2O_3 content reaches 10 m/O or more the phase relation has been greatly changed. The ξ-phase always appears by cooling an $LiIO_3$ melt containing γ-Al_2O_3 particles to room temperature. This means that a suitable amount of γ-Al_2O_3 tends to stabilize the ξ-phase. Figure 4 shows the phase relation for both pure $LiIO_3$ and $LiIO_3$ (0 m/O γ-Al_2O_3) systems.

α-$LiIO_3$ has a rather strong electric polarity resulting from parallel arrangement of IO_3^- polar groups. β-$LiIO_3$ has central symmetry. In its unit cell the polar groups are arranged antiparallel to each other. In γ- and ξ-phases IO_3^- polar groups are neither parallel nor antiparallel, but in medium status. When $LiIO_3$ changes from α- to γ- and then to β-phase, the IO_3^- ionic groups must rotate to reach the required configurations. Thus, the increase or decrease of phase transition temperature indicates the degree of difficulty for the rotational movement of IO_3^- groups. At high pressure the phase transition temperature for α- to β-phase increases with applied pressure. This means that applied pressure impedes the rotational movement of IO_3^- groups. In a dispersion system, the matrix phase $LiIO_3$ may be strained due to the difference of expansion between $LiIO_3$ and γ-Al_2O_3 during thermal cycling. Similarly, with high pressure the interaction between the IO_3^- group and strain field would result in an increase of phase transition temperature. On the other hand, the fact that the high pressure ξ-phase could be obtained rather easily by cooling $LiIO_3$ melt containing γ-Al_2O_3 particles may also be an indication of the similarity between the behavior of $LiIO_3$ at high pressure and at dispersion state. Therefore both the increase of the phase transition temperature and the appearance of the ξ-phase may be due to the same interaction between IO_3^- groups and the strain field.

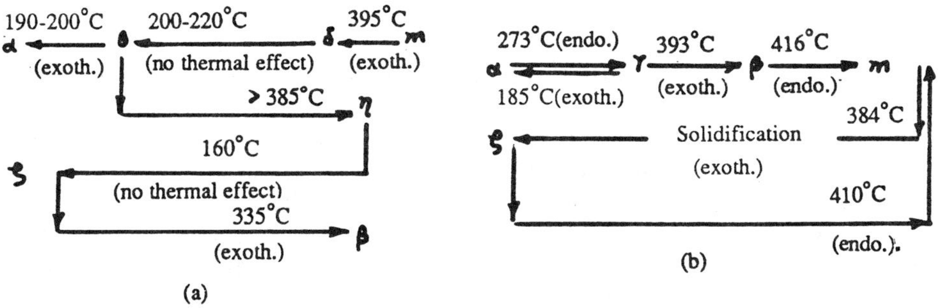

Fig. 4. Phase relation for pure $LiIO_3$ (a) and $LiIO_3$ (10 m/O γ-Al_2O_3) (b).

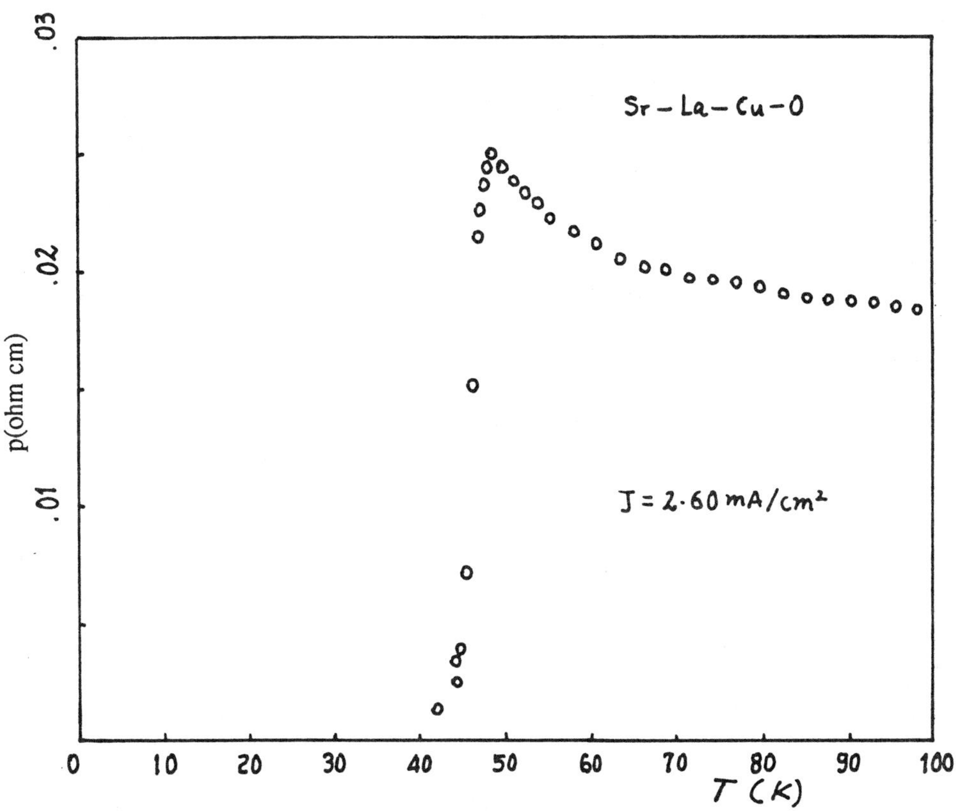

Fig. 5. The temperature dependence of resistivity for $Sr_{0.5}La_{4.5}Cu_5O_{5(3-y)}$.

4. Superconductivity in Multiphase Oxides

Recently Bednorz and Muller [11] proposed that a new oxide $La_{5-x}Ba_xCu_xO_{5(3-y)}$ could be a possible high T_c superconductor. Their electrical measurements showed that the superconducting transition temperature T_c might be as high as 35 K, 12 K higher than that for Nb_3Ge. The samples prepared by them are multiphase, i.e., a dispersion system. The drop was found to shift towards lower temperature as the measuring current was increased. Possible high-temperature percolative superconductivity with a maximum onset temperature of T_{co} = 35 K was therefore proposed. This is one of the reasons why we are interested in these kinds of oxide superconductors.

We have studied superconductivity in both Ba-La-Cu-O and Sr-La-Cu-O systems. Among dozens of samples with nominal composition $(Sr(Ba)_{0.5}La_{4.5}Cu_5O_{5(3-y)}$ the highest T_{co} for Sr-La-Cu-O is 48.6 K with a transition width of 10 K (Figure 5). In the Ba-La-Cu-O system the highest T_{co} is 46.3 K with a transition width of 7 K (Figure 6). These T_{co} values were the highest recorded in the world at that time. More importantly, we have observed superconductivity phenomenon at 70 K with these samples many times [12]. However, the superconductivity for both these systems is not stable. The T_{co} decreases from 48.6 K to 35.05 K for Sr-La-Cu-O after 10 days. For Ba-La-Cu-O a the T_{co} dropped from 46.3 K to 30.4 K after 4 days.

Cava et al. [13] obtained a nearly single superconducting phase with T_c of 36.2 K; Ushida et al. [14] and Takagi et al. [15] have identified the superconducting phase as $(Ba_xLa_{1-x})_2CuO_4$ with a K_2NiF_2 layer — perovskite structure. Most researchers have accepted the above conclusion and have obtained good results with T_c's of approximately 30-40 K. WE have also synthesized samples with this composition. The T_{co} is ~44 K in a Ba-La-Cu-O system.

We have substituted Y for La to elevate the superconducting transition temperature. The experimental results show that the midpoint of the transition determined by resistivity is 92.8 K with a transition width of 4 K and the onset of the transition is above 100 K. The temperature dependence of resistance departs from linear behavior at about 110 K (Figure 7). At 78.5 K the resistance approaches zero. During the transition the resistance decreases by five orders of magnitude. A. C. magnetic susceptibility measurements show that diamagnetism begins at 93 K, which corresponds to the midpoint of the resistivity transition. By comparison to the voltage level of the standard Pb with the same volume, at 77 K, the fraction of the superconducting phase should be about 13% if the sample is a single-connected region. X-ray diffraction patterns show that the samples are multiphase.

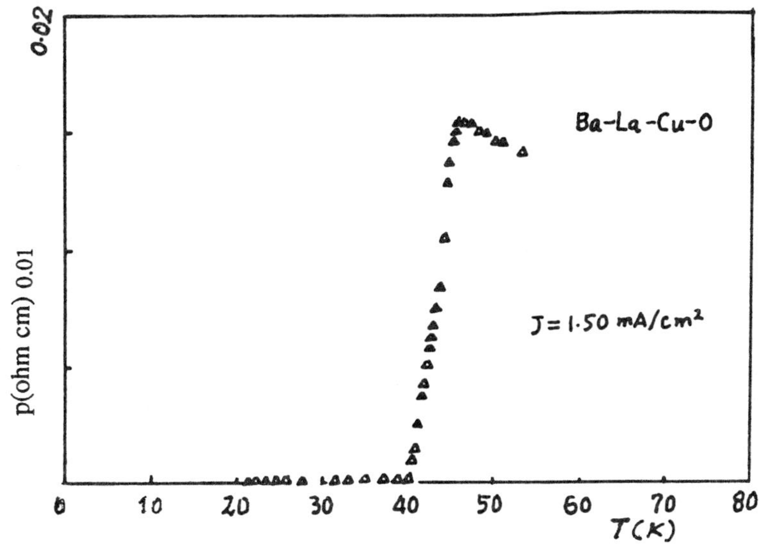

Fig. 6. Temperature dependence of resistivity for $Ba_{0.5}La_{4.5}Cu_5O_{5(3-y)}$.

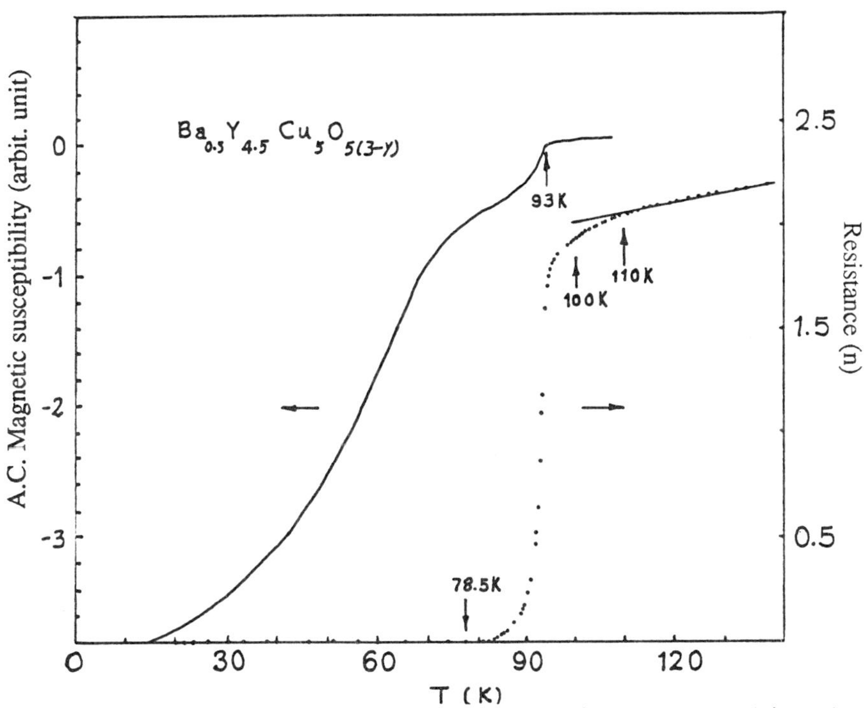

Fig. 7. Temperature dependence of resistivity (right-hand scale) and A.C. magnetic susceptibility (left-hand scale) of sample $Y_{4.5}Ba_{0.5}Cu_5O_{5(3-y)}$.

The single superconducting phase in this system is $YBa_2Cu_3O_{9-x}$. For single-phase the resistance drops to zero at about 92 K more rapidly.

More recently [16] we have prepared a series of new superconducting oxides Ba-Re-Cu-O, where Re = Lu, Yb, Er, Eu... The experimental results show that the midpoint of the transition determined by resistivity is 90.5 K for Lu and 90.8 K for Yb with a transition width of 3 K for Lu and 5 K for Yb. The series of Lu's fourfold element oxides Ba-Lu-M-Cu-O with M = Dy, La, Y and the series of Ba-Y-D-Cu-O, based on Ba-Y-Cu-O, with D = Dy, Lu have also been obtained as new high T_c superconductors.

From the above results, we believe that although the sample with nominal composition of $Ba_{0.5}Re_{4.5}Cu_5O_{5(3-y)}$ will be multiphase, it is beneficial to form the high T_c phase. On the other hand, interfacial effects appear to play some role in superconductivity.

References

1. Liquan, C., In: *Materials for Solid State Batteries*, Ed. by B. V. R. Chowdari and S. Radhakrishna, World Scientific, p. 69, 1986.
2. Liquan, Chen et al. *Acta Physica*, vol. 35, p. 1158, 1986.
3. Phase Diagram and Phase Transition Research Group, *Acta Physica*, vol. 26, p. 199, 1977.
4. Xu jian et al. *Acta Physica*, vol. 29, p. 1065, 1980.
5. Zhongyi, Shen, Yun, Shung, *Kexue Tongbao*, vol. 26, p. 913, 1980.
6. Yushu, Yao et al. *Acta Physica Sinica*, vol. 30, p. 835, 1981.
7. Jun, Liu et al., *Kexue Tongbao*, vol. 27, p. 979, 1982.
8. Zhongyi, Shen et al., *Acta Physica Sinica*, vol. 32, p. 118, 1983.
9. Jingkui, Liang, Yumin, Zhang, *The Journal of Structure Chemistry*, vol. 2, p. 81, 1983.
10. Jingkui, Liang, Yumin, Zhang, *Acta Physica*, vol. 33, p. 69, 1984.
11. Bednorz, J. G., Muller, K. A., *Z Phys. B*, vol. 64, p. 189, 1986.
12. Zhongxian, Zago, Liquan, Chen, et al., *Kexure Tongbao*, no. 6, 1987.
13. Cava, R. J. et al., *Phys. Rev. Lett.*, vol. 58, p. 408, 1987.
14. Uchida, S. et al., *Jpn. J. Appl. Phys. Lett.*, in press.
15. Takagi, H. et al., *Jpn. J. Appl. Phys. Lett.*, in press.
16. Zhongxian, Zhao, Liquan, Chen, et al., *Kexure Tongbao*, in press.

POLYMER SOLID ELECTROLYTES

B. E. Mellander
Department of Physics
Chalmers University of Technology
S-412 96 Göteborg, Sweden

Abstract: **During the last few years a number of ion conducting polymers have been discovered and conductivities of the order of 10^{-3} (ohm cm)$^{-1}$ have been reported. Ion conducting materials are promising as the electrolyte in different types of power sources.**

1. Introduction

Polymeric materials are usually very good electrical insulators with conductivities as low as 10^{-18} (ohm cm)$^{-1}$. However, during the last few years both electron and ion conducting polymeric materials have been the subject of a rapidly increasing number of scientific studies (see, for example, Ref. [1-5] for ion conducting polymers and [6] for electron conducting polymers). In the present paper the research on ion conducting polymers will be reviewed.

In the first reports on ion conducting polymer materials poly(ethylene) oxide (PEO) and poly(propylene) oxide (PPO)-alkali metal salt systems were studied [1, 7, 8]. PEO and PPO have simple repeat units (see Figure 1) and the degree of polymerization for PEO can be higher than 100,000 corresponding to a molecular weight of five million. In addition to PEO and PPO, a number of more complicated polymers are now being used. The salts used have been mainly lithium salts such as LiI, LiSCN, $LiClO_4$, $LiCF_3SO_3$, and the corresponding sodium salts. However, potassium, cesium, and silver salts have also been used in some studies.

2. Sample Preparation

A common preparation method is to dissolve the polymer and salt in a common solvent such as acetonitrile or methanol and then evaporate off the solvent. A number of other methods have also been used, for example, depositing thin films by vacuum evaporation [9]. For all preparation methods, utmost care must be taken to avoid any contact with water, since polymer, salt, and solvent are very often hygroscopic. The polymer to salt concentration is often given as the number of oxygen atoms in the polymer chain per cation, the O:M ration because the unshared electron pair of the oxygen atom is important for the solvation of cations.

3. Ion Conductivity

The ionic conductivity of polymer electrolytes is usually determined using complex impedance spectroscopy [10]. In Table 1 the ionic conductivity at 85°C is shown for some ion conducting polymers. The ionic conductivity for most of these materials falls rapidly at lower temperatures and is often of the order of 10^{-7} (ohm cm)$^{-1}$ or lower at room temperature. For crystalline materials the temperature dependence of the ionic conductivity can be described by

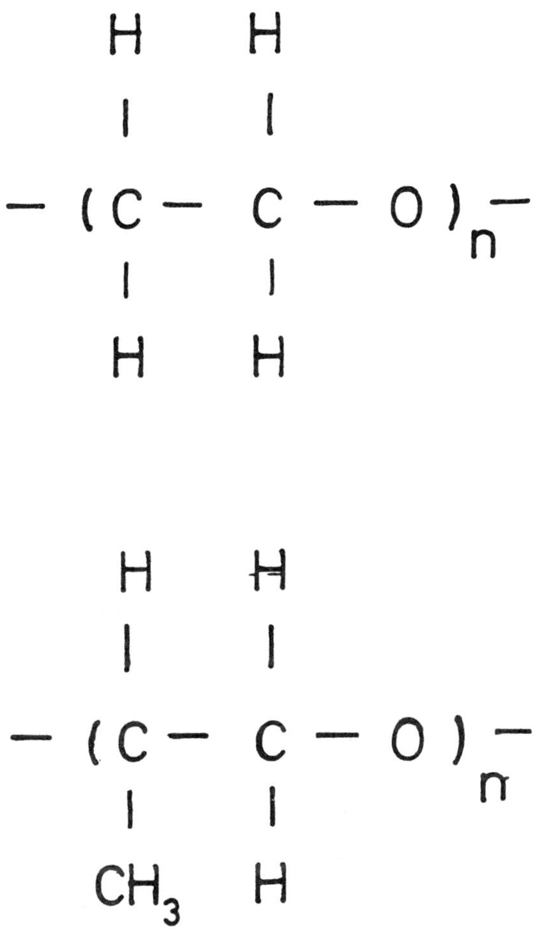

PEO

PPO

Fig. 1. Repeat units of poly(ethylene oxide) (PEO) and poly(propylene oxide) (PPO) where n is the degree of polymerization.

Table 1

Ionic conductivities at 85°C for some ion conducting polymers based on poly(ethylene) oxide.

	Conductivity $(ohm\ cm)^{-1}$	Reference
$P(EO)_{100}LiCF_3SO_3$	$3\ 10^{-4}$	11
$P(EO)_{12}LiClO_4$	$1\ 10^{-3}$	11
$P(EO)_{20}LiAsF_6$	$1\ 10^{-3}$	11
$P(EO)_{4.5}LiI$	$9\ 10^{-6}$	12
$P(EO)_{4.5}LiSCN$	$3\ 10^{-5}$	12

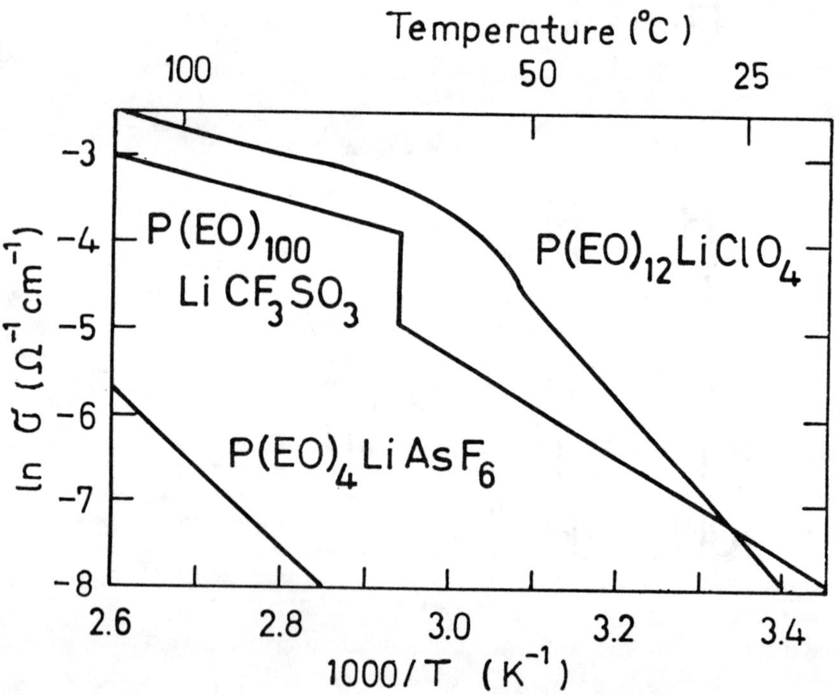

Fig. 2. Ionic conductivities for some ion conducting polymers based on poly(ethylene oxide). The conductivity values are taken from Ref. [11].

$$\sigma T = (\sigma T)_0 \exp(-\Delta H/kT) \quad (1)$$

where σ is the ionic conductivity, T is the absolute temperature, ΔH is the activation enthalpy, k is Boltzmann's constant and $(\sigma T)_0$ is a pre-exponential factor. For amorphous materials the ionic conductivity can instead by described by the Vogel-Tammann-Fulcher (VTF) equation [13-15]:

$$\sigma = AT^{-1/2}\exp(-E/(T - T_0)) \quad (2)$$

where A is a pre-exponential factor, E is pseudoactivation energy and T_0 is the ideal glass transition temperature. In a $\ln(\sigma T)$ versus $1/T$ plot Equation (2) corresponds to a curve while Equation (1) corresponds to a straight line. The temperature dependence of the ionic conductivity for some polymer solid electrolytes are shown in Figure 2 where both types of behavior can be seen. For $P(EO)_{12}LiClO_4$ the conductivity curve is VTF-type at higher temperatures and straight line-type at lower temperatures, while for $P(EO)_{100}LiCF_3SO_3$ two straight line portions appear on both sides of a transition.

The ionic conductivity of poly(ethylene oxide)-lithium salt systems have been extensively studied recently, especially systems containing LiI, $LiClO_4$, LiSCN, and $LiCF_3SO_3$ (see, for example, [1, 11, 12, 16]). In most cases PEO is partly crystalline and it has been shown that the ion transport takes place mainly in the amorphous regions in the material while crystalline regions have a lower conductivity [17]. In general, highly symmetric polymers such as PEO (Figure 1) have a much higher degree of crystallinity than polymers with large side groups such as PPO or poly(NN-dimethyl acryl amid) (PDMAA) (Figure 3). PDMAA has a very large side group which also includes the oxygen and forms a completely amorphous polymer. In order to obtain higher ionic conductivities amorphous materials containing PPO as well as systems with other polymer hosts have been investigated and it has been shown that the flexibility of the polymer backbone and sidegroups are important for ion mobility [18, 19].

The ion conducting materials that have been studied are not so often single phase materials. Instead two or more phases are present, for example, pure crystalline or amorphous polymer, polymer-salt complexes, and pure salt phases. It was first supposed that a 0:M ratio of about 4 would give the highest conductivities but it has now been shown that conductivity maxima may be obtained for very low salt concentrations, for example, O:Li = 100 in the P(EO) - $LiCF_3SO_3$ system [11]. In order to study the composition dependence of the ionic conductivity, phase diagrams have been introduced [11, 20-22]. Materials with high salt concentrations have also been studied where a good solid electrolyte is mixed with poly(ethylene oxide) [23].

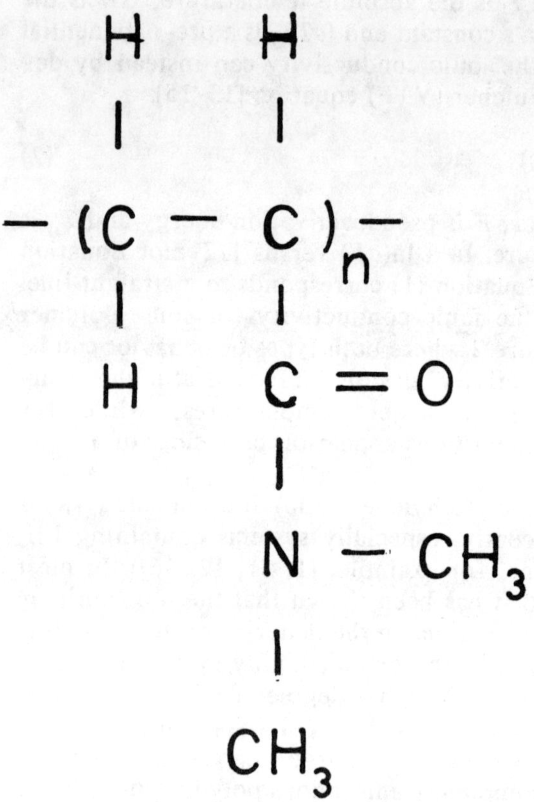

PDMAA

Fig. 3. Repeat unit of poly(NN-dimethyl acryl amid).

For the use of a solid electrolyte in a battery application electronic conductivity must be very low since an electronic charge transport would cause internal discharging of the cell and it has been shown for different polymers that the electronic conductivity in ion conducting polymers is negligible [19, 24]. Another requirement is that only cations (or anions) are mobile in the electrolyte. At first it was supposed that the polymeric solid electrolytes were pure cation conductors [1] but several measurements of the cation transport number t_+, i.e., the fraction of the current that is transported by cations, have shown that both cations and anions are mobile. Using impedance spectroscopy and potentiometric measurements the cation transport number has been determined to be 0.54 for $P(EO)_{4.5}LiSCN$ [25], 0.3 for PEO-LiI, and PEO-$LiClO_4$, and 0.7 for PEO-$LiCF_3SO_3$ [26]. Measurements on a number of other ion conducting polymers have also shown that often both anions and cations are mobile [17, 27-29].

4. Applications

The obvious application of ion conducting polymers is as the electrolyte in different types of power sources, and rechargeable batteries using polymer electrolytes have been demonstrated [30]. Ion conducting polymers have many advantages compared to the classic solid electrolytes due to the plastic properties of the polymer. A polymer solid electrolyte can, for example, easily absorb volume changes at the electrode-electrolyte interfaces which occur during battery operation. It is possible to produce ion conducting thin films which somewhat reduces the requirement of very high ionic conductivity for a battery electrolyte. Furthermore, mass production of polymer materials is, in general, easy, and the materials are light weight and have a relatively low price. However, there are still some disadvantages with polymer solid electrolytes, for example, low conductivity at room temperature, and the fact that both cations and anions are mobile.

References

1. Armand, M., Chabagno, J., Duclot, M., In: *Fast Ion Transport in Solids*, Ed. by. P. Vashishta, J. N. Mundy, G. K. Shenoy, New York: North Holland Publishing Co., p. 131, 1979.
2. Shriver, D. F., Papke, B. L., Ratner, M. A., Dupon, R., Wong, T., Brodwin, M., *Solid State Ionics*, vol. 5, p. 83, 1981.
3. Armand, M., *Solid State Ionics*, vol. 9, 10, p. 745, 1983.
4. Davis, G. T., Chiang, C. K., In: *Conducting Polymeric Materials*, Ed. by H. Sasabe, Tokyo: Chemical Marketing Co., p. 244, 1984.
5. Armand, M., *Ann. Rev. Mater. Sci.*, vol. 16, p. 245, 1986.
6. Bott, D., *Phys. Technol*, vol. 16, p. 121, 1985.
7. Fenton, D. E., Parker, J. M., Wright, P. V., *Polymer*, vol. 14, p. 589, 1973.

8. Wright, P. V., *Br. Polymer J.*, vol. 7, p. 319, 1975.
9. Ito, Y., Kanehori, K., Miyauchi, K., Kudo, T., *Solid State Ionics*, vol. 23, p. 69, 1987.
10. Mellander, B. E., Lunden, A., In: *Materials for Solid State Batteries*, Ed. by. B. V. R. Chowdari, S. Radhakrishna, Singapore: World Scientific Publ. Co., p. 161, 1986.
11. Robitaille, C. D., Fauteux, D., *J. Electrochem. Soc.*, vol. 133, p. 315, 1986.
12. Rietman, E. A., Kaplan, M. L., Cava, R. J., *Solid State Ionics*, vol. 17, p. 67, 1985.
13. Vogel, H., *Phys. Z*, vol. 22, p. 645, 1921.
14. Tammann, V. G., Hesse, W., *Z. Anorg. Allg. Chem.*, vol. 156, p. 245, 1926.
15. Fulcher, G. S., *J. Am. Ceram. Soc.*, vol. 8, p. 339, 1925.
16. Weston, J. E., Steele, B. C. H., *Solid State Ionics*, vol. 7, p. 81, 1982.
17. Berthier, C., Gorecki, W. Minier, M., Armand, M. B., Chabagno, J. M., Rigaud, P., *Solid State Ionics*, vol. 11, p. 91, 1983.
18. Fontanella, M. C., Wintersgill, M. C., Smith, M. K., Semancik, J., Andeen, C. G., *J. Appl. Phys.*, vol. 60, p. 2665, 1986.
19. Blonsky, P. M., Shriver, D. F., Austin, P., Allcock, H. R., *Solid State Ionics*, vol. 18, 19, p. 258, 1986.
20. Sorensen, P. R., Jacobsen, T., *Polymer Bull.*, vol. 9, p. 47, 1983.
21. Minier, M., Berthier, C., Gorecki, W., *J. Physique*, vol. 45, p. 739, 1984.
22. Lee, Y. L., Crist, B., *J. Appl. Phys.*, vol. 60, p. 2683, 1986.
23. Stevens, J. R., Mellander, B. E., *Solid State Ionics*, vol. 21, p. 203, 1986.
24. Dupon, R., Whitmore, D. H., Shriver, D. F., *J. Electrochem. Soc.*, vol. 128, p. 715, 1981.
25. Sorensen, P. R., Jacobsen, T., *Electrochim. Acta*, vol. 27, p. 1671, 1982.
26. Bouridah, A., Dalard, F., Deroo, D., Armand, M. B., *Solid State Ionics*, vol. 18, 19, p. 287, 1986.
27. Leveque, M., le Nest, J. F., Gandini, A., Cheradame, H., *J. Power Sources*, vol. 14, p. 27, 1985.
28. Ferloni, P., Chiodelli, G., Magistris, A., Sanesi, M., *Solid State Ionics*, vol. 18, 19, p. 265, 1986.
29. Bhattacharja, S., Smoot, S. W., Whitmore, D. H., *Solid State Ionics*, vol. 18, 19, p. 306, 1986.
30. Hooper, A., North, J. M., *Solid State Ionics*, vol. 9, 10, p. 1161, 1983.

SOLID STATE LITHIUM ION CONDUCTORS FOR MODERATE TEMPERATURES

Steen Skaarup
Physics Laboratory III
Technical University of Denmark, DK-2800
Lyngby, Denmark

Abstract: **The lithium ion occupies a unique position among the alkali metal ions because of its small size (0.06 Å). The small ionic radius has important consequences both for the stability of ionic compounds (i.e., lithium nitride is the only stable alkali metal nitride) and for the existence of a number of crystalline compounds with appreciable ionic conductivity at temperatures below the melting point of lithium (180#C). In addition to the crystalline compounds, lithium ion conductivity has also been found in glasses, composite electrolytes and in organic polymers.**

Selected moderate temperature Li^+ ion conductors are discussed with respect to the widely varying structures and mechanisms for ionic motion, as well as to the thermodynamic, interfacial, and mechanical properties determining the chemical stability and the ease of fabrication. Experimental data are presented for mixed phase electrolytes which represent and attempt to combine the advantages of crystalline and polymer electrolytes.

Introduction:

All ionic compounds show some degree of ionic mobility, but the ionic conductivity is usually extremely low. This is because the strong repulsive potential between two ions at close distances, where the electron clouds begin to overlap appreciably, makes the ions behave essentially as hard spheres with little possibility of passing each other in the crystal structure. Ionic structures are usually quite densely packed with ions in order to maximise the stabilizing Coulomb forces (Madelung potential). The reason for the nonzero conductivity observed is the presence of defects, such as vacancies or interstitial ions which are present in small amounts even in thermodynamic equilibrium |1|. The defects make conduction possible, but as seen from Table 1, the conductivities are far lower than those obtained in aqueous solutions and molten salts ($\sigma = 0.1 - 1\ \Omega^{-1}cm^{-1}$).

Table 1

Ionic Conductivity of Lithium Halides at Room Temperature |2|

salt	σ_{25} ($\Omega^{-1}cm^{-1}$)
LiF	10^{-13}
LiCl	$5*10^{-10}$
LiBr	$2*10^{-9}$
LiI	$1*10^{-7}$

A small fraction of ionic compounds show much higher ionic conductivity - usually both because they contain a large number of defects (either intrinsic to the structure or created by doping) and because the lattice has special conduction paths with low activation energy. These so-called fast ion conductors reach ionic conductivities below their melting points comparable to those of molten salts and present new opportunities for making solid state electrochemical devices.

Fast ionic conduction has been found for a range of mostly monovalent cations. The Ag^+ conductors based on AgI are the best ionic conductors known, but technological interest has been centered around alkali metal conductors, especially Li and Na, because the highly electropositive metals make high energy density batteries possible, and for economic and resource reasons.

This article deals with Li^+ ion conductors, and further restricts the discussion to systems reaching technologically interesting conductivities ($\sigma = 10^{-4} - 10^{-3}\ \Omega^{-1}cm^{-1}$) below the melting point of lithium (180 °C).

Moderate Temperature Li^{+} Conductors:

Table 2 contains the conductivity data for four of the best Li^{+} conductors in the temperature range below 180 °C. Together they represent most of the types of presently known fast ionic conductors. The room temperature conductivities exceed those of the simple salts by many orders of magnitude. They all belong to the group of class I (or normal melting) ionic conductors |7|, this is in contrast to some Ag^{+} and Cu^{+} salts that undergo a phase transition below the melting point accompanied by a drastic increase in ionic conductivity (class II compounds).

The structural aspects that cause the high ionic conductivity as well as the mechanisms of conduction vary widely in the series which therefore illustrates the variety found in fast ionic conduction.

Table 2

Ionic conductivity data for selected Li^{+} solid state ionic conductors

compound	σ_{25} ($\Omega^{-1}cm^{-1}$)	σ_{150}	E_a (eV)	type	reference
Li_3N+1%H	$6*10^{-3}$	$4*10^{-2}$	0.20	doped	|3|
Li_3N	$3*10^{-4}$	$4*10^{-3}$	0.26	intrinsic	|3|
LiI/Al_2O_3	$4*10^{-5}$	$2*10^{-3}$	0.36	composite	|4|
$Li_2S-B_2S_3-LiI$	$2*10^{-3}$	$(4*10^{-2})$[1]	0.30	glass	|5|
$LiCF_3SO_3(PEO)_{10}$	$3*10^{-8}$	$1*10^{-3}$	1.25[2];0.62[3]	polymer	|6|

[1]Unstable above ~80 °C. [2]Below ~70 °C. [3]Above ~70 °C.

Conduction Mechanisms:

Hydrogen doped Li_3N has the highest room temperature conductivity of any Li^{+} ionic conductor. When a current of lithium is passed through the crystal, the hydrogen is slowly removed and the conductivity decreases to that of the pure compound which is smaller by a factor of 20 |8|. Pure Li_3N is still one of the best crystalline Li^{+} ion conductors (only surpassed by lithium beta alumina with $\sigma_{25} = 3*10^{-3}\ \Omega^{-1}cm^{-1}$, E_a = 0.24 eV |9|).

The crystal structure of Li_3N is shown in figure 1. The hexagonal structure, which is unique to this compound, is characterised by layers of N^{3-} ions surrounded by 6 Li^+ ions in the layer and 2 bridging Li^+ ions. The reason for the high ionic mobility is not that the ions move in the large tunnels between layers evident in the structure, but the charge is transported by the motion of vacancies in the layers. The vacancies are present even in the pure, stoichiometric form because the large cavities easily accomodate an interstitial Li^+ ion - the energy of formation of a Frenkel defect is only about 0.1 eV |10|. The effect of substituting hydrogen for lithium in the layers (maximum: About 1 atom%) is to weaken the neighbouring Li-N bonds thus lowering the energies of Frenkel defect formation and of vacancy migration. The doping enhances the already existing intrinsic mechanism by lowering the activation energy.

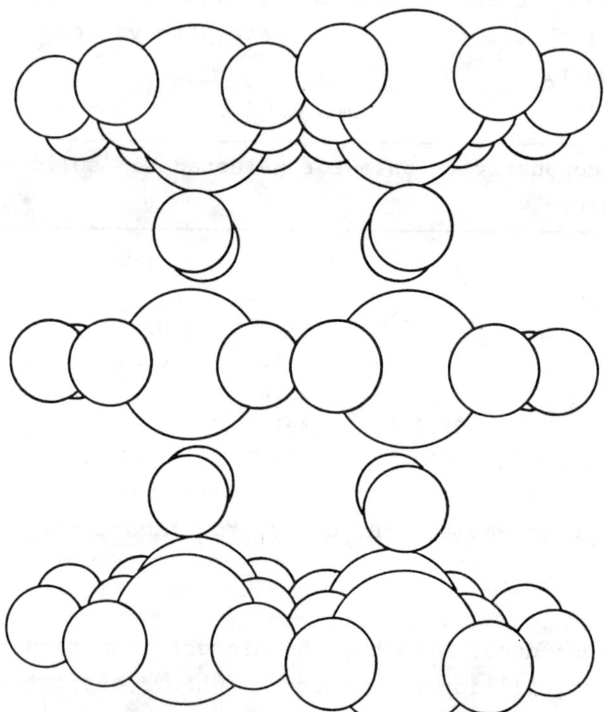

Figure 1. The structure of Li_3N. Hexagonal layers of Li^+ (small spheres) and N^{3-} ions are connected by bridging Li^+ ions.

The mechanism of ionic conduction in the LiI/Al_2O_3 composite is very different and is still not completely understood. LiI itself has only a very limited ionic conductivity (Table 1), but the addition of non-conducting Al_2O_3 of high surface area increases the room temperature conductivity by about three orders of magnitude. The resulting mixture is termed a composite electrolyte, and several theoretical models have been proposed in order to explain the phenomenon, which has been observed in a number of combinations of electrolytes and high surface area components, although the enhancement factors are usually a lot smaller than in the case of LiI/Al_2O_3. The conductivity enhancement increases with the specific surface area of the alumina powder pointing towards surface rather than bulk conduction. A plausible mechanism is the creation of a large number of mobile Li^+ vacancies in the lithium iodide closest to the alumina surface either because of space charge effects or because of adsorption of Li^+ ions at nucleophilic surface groups (e.g. OH) |11|. The conductivity enhancement is characterised by going through a maximum as a function of the amount of alumina (figure 2) - this can be explained as the result of two percolation limits: At low volume-percentages there is too little Al_2O_3, and at high percentages too little LiI to form unbroken interfacial ionically conducting paths through the mixture.

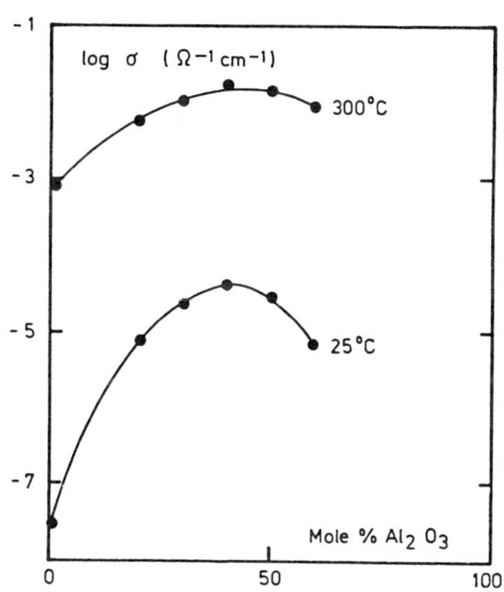

Figure 2. Ionic conductivity of LiI/Al_2O_3 composite as a function of the amount of alumina in the electrolyte mixture.

Since ionic conduction depends on the presence of defects and structural disorder, glasses might be thought to be ideal candidates for solid electrolytes. Normal silicate glasses show very little ionic conductivity, however, since the migration energy required to move in the strongly bonded Si-O backbone is too high. Glasses of much higher ionic conductivity have been found based on B_2S_3 as the glass former. The data in Table 2 are for 0.30/0.26/0.44 mole % $Li_2S/B_2S_3/LiI$.

The mechanism of conduction is the motion of Li^+ ions in the immobile B-S-I backbone. The number of potentially mobile ions is very high in a glass, and because the large S^{--} and I^- ions are very polarizable, in contrast to O^{--}, a high degree of ionic mobility is also possible.

The ionically conducting organic polymers containing lithium salt (most are based on poly ethylene oxide PEO, repeating unit: $-CH_2-CH_2-O-$), form a quite different class of ionic conductors from the 3 inorganic systems mentioned above. As seen from the Arrhenius plot of the ionic conductivities in figure 3, the conductivity is characterized by two activation energies - a high value below ~70 ^oC and a lower value above this transition temperature.

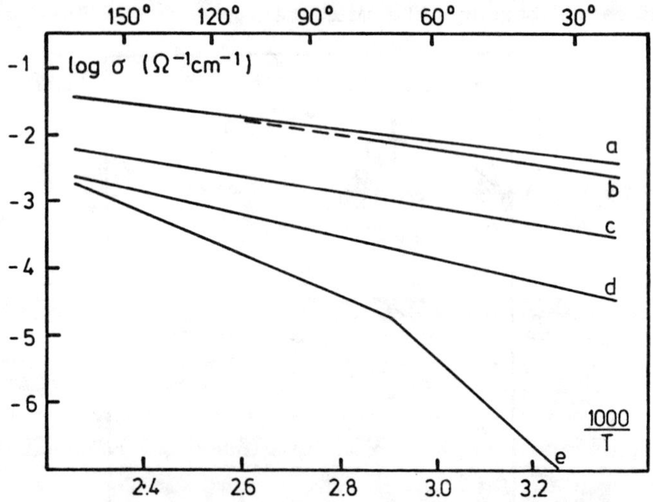

Figure 3. Variation of ionic conductivity with temperature for selected Li^+ ion conductors.

a) H-doped Li_3N, b) $Li_2S-B_2S_3-LiI$ glass, c) Intrinsic Li_3N, d) LiI/Al_2O_3 composite, e) $LiCF_3SO_3(PEO)_{10}$ polymer.

When first discovered, the ionic motion in polymers was believed to take place in the crystalline regions and to be due to the Li^+ ions solely - the process was regarded as analogous to the conduction process in crystals. It was later realized, that the motion of ions takes place primarily in the amorphous regions and that both cations and anions move (as in liquid electrolytes) - a fact which has important consequences for the usefulness as electrolyte in a battery |12|. The ionic motion is made possible at higher temperatures because of fast fluctuating motions of the polymer chains which create temporary paths for ions hopping. Below about 70 ^{o}C the chain motions become much slower as the crystalline regions grow, and the ionic mobility decreases rapidly. Attempts at further improving the conductivity therefore concentrate on stabilizing the amorphous regions to lower temperatures, for instance by modifying the backbone of the polymer.

Stability and Interfacial Properties:

Ionic conductivity is only one of the parameters that determine the usefulness of solid electrolytes in electrochemical devices. One of the crucial processes in solid state electrochemical systems is the transfer of ions across the interfacial boundary between two phases. Good interfacial contact is more difficult to achieve between solid phases than if one phase is liquid.

Of equal importance is the chemical and electrochemical stability when in contact with other components of the system. Li_3N, Li_2S-B_2S_3-LiI glass and LiI/Al_2O_3 composite are all chemically stable with respect to metallic lithium. This yields the advantage of good cyclability of lithium across the metal/electrolyte interface since no new phase is formed which might block the ion transfer. The main limiting factor for the current density will be the speed at which lithium atoms from the bulk of the metal can move to fill the voids created at the interface. The selfdiffusion in lithium is not large enough to quarantee sufficient replenishment of the interfacial contact at moderate temperatures and current densities of the order of 1 mA/cm^2 |13|. This will often necessitate the use of pressure on the system, for instance by incorporating spring loading in the design. At temperatures above ~100 °C, only small pressures are needed because of the softness of lithium metal, but at temperatures closer to room temperature, this may turn out to be a serious limiting factor.

As an organic, oxygen containing system, $LiCF_3SO_3(PEO)_{10}$ probably is not thermodynamically stable against metallic lithium, but the polymer is practically stable for at least several months. In the longer term, a layer of probably LiF builds up and retards the ion transfer |14|. Other stable lithium salt layers are formed when using e.g. LiI or $LiClO_4$ in PEO. If the reaction layer is thin, its conductivity does not have to be large for the resistance to be negligible. However, the long term stability of any organic system in contact with lithium at temperatures of the order of 100 °C is still somewhat in doubt.

Similar considerations apply to the interface formed by the electrolyte and the positive electrode. LiI is limited to potentials less than about 2.8 V with respect to lithium because of the oxidation of iodide to free iodine. This is also the practical upper voltage limit of LiI/Al_2O_3 and of LiI in poly ethylene oxide. In the lithium iodide containing glass $Li_2S\text{-}B_2S_3\text{-}LiI$, the practical stability is reported to be 4.5 V. This is explained by the iodine atoms being so tightly bound in the glass matrix that they are immobile and unable to recombine. The upper thermodynamic stability limit of Li_3N (corresponding to the release of free nitrogen) is only about 0.44 V which is too low to be interesting for battery use. The kinetics of the reaction seems to be so slow, however, that the practical limit appears to be about 2.5 V |15|. The slow rate of recombination may also in this case be due to the immobility of the tightly bound N^{3-} which are surrounded by 8 Li^+ ions.

Mechanical Properties and Ease of Fabrication:

When working with lithium ionic compounds it is usually necessary to take precautions against moisture since the Li^+ ion is very hygroscopic due to its small size. All four electrolytes treated here are hygroscopic and some will require dry room or even glove box facilities for handling. The polymer electrolyte has the advantage that it can be made in ambient atmosphere because absorbed water can be removed reversibly by heating to about 120 °C in vacuum. The high surface area Al_2O_3 used in the composite is very hygroscopic and can only be regenerated by heating in vacuum above 600 °C. Li_3N is quickly covered by a protective reaction layer of LiOH and Li_2CO_3 when exposed to air. When working with large samples, this layer can be removed by abrasion, but in the case of powder with high surface area this is impossible, and the pure Li_3N surface cannot be reestablished. The glass electrolyte demands the highest degree of protection against moisture - at H_2O contents above about 5 ppm, H_2S is irreversibly formed and the surface is progressively destroyed. The necessity of working under very pure atmospheres may well be one of the factors limiting the technological applicability.

The mechanical properties vary from the very hard and brittle Li_3N over the somewhat softer composite and glass to the very soft polymer. Small (~1 cm^2) compact pellets of Li_3N can be made by sintering or hot pressing under controlled atmosphere. Glass samples of about the same size can be fabricated by rapid quenching from about 800 °C (the glass is not stable above ~80 °C because of devitrification). LiI/Al_2O_3 pellets of about 90% theoretical density can be pressed at room temperature. These methods suffice for small area applications, but it is difficult to see how the three inorganic materials could be easily manufactured in larger areas.

Larger sheets of Li_3N or glass would probably also readily crack because of structural changes due to chemical reactions or thermal stresses, whereas LiI/Al_2O_3 might have insufficient structural stability. The polymer electrolyte is ideal in this respect since it can easily be manufactured in large sheets, and since the technology for the continuous fabrication of polymer sheets already exists (e.g. in the manufacture of photographic film). The softness and flexibility of the polymer is a further advantage since it enables the electrolyte to maintain contact to the other solid phases during structural changes.

Mixed Phase Electrolytes:

The comparison of the moderate temperature Li^+ ion conductors above shows that none of them emerges as obviously superior The polymer has clear advantages with respect to achieving and retaining solid/solid contact and with respect to ease of fabrication, but the room temperature conductivity is much lower than those of the inorganic conductors. Furthermore, the fact that the anions move as well as the Li^+ ions limits the current density unless extremely thin layers can be used. Mixed phase electrolytes represent an attempt to combine the advantages of the polymer with the higher conductivity and unity Li^+ transport number of the inorganic conductors.

Figure 4 shows a schematic representation of an ideal mixed phase electrolyte. The structure consists of a large volume percentage (80-95%) of crystalline or glassy ionic conductor. The grains are "glued" together by thin coatings of polymer electrolyte that fill the intergranular space. This takes advantage of the excellent contact properties and flexibility of the polymer while the ionic conduction mostly takes place in the grains. The polymer layers should be so thin that their contribution to the resistivity is negligible and that no limiting current effects appear. The approximate average thickness of the polymer layer is equal to $(1-v)r/3v$ where v is the volume fraction of crystalline grains of radius r. For $v>0.8$ and $r>50$ μm, the typical thickness is smaller than 4 μm.

Figure 4. Schematic representation of a mixed phase electrolyte. White: Crystalline ionic conductor. Black: Polymer electrolyte.

The differences between a mixed phase electrolyte and a composite electrolyte are: Since the aim is to take advantage of the excellent bulk ionic conductivity of the major component, the grains should be rather large (50-100 μm) to avoid being limited by interfacial conduction, whereas the conduction in the composite electrolyte is an interfacial process and therefore requires very small grain sizes (corresponding to surface areas of 100-200 m^2/g). The proportion of the minor component is kept as small as possible compatible with retaining the desired mechanical properties, whereas the composite electrolyte is a mixture of about 50/50 volume percent.

The ideal mixed phase electrolyte would have an activation energy for the conductivity equal to that of the major component, since the ionic motion in the polymer and along or across the interfaces are assumed not to be limiting.

Figure 5 shows the resulting conductivity of mixed phase electrolytes made from $LiCF_3SO_3(PEO)_{10}$ polymer combined with Li_3N and Li_2S-B_2S_3-LiI glass respectively. The samples have been made by mixing and pressing the powders with about 10 ton/cm^2 at about 60 ^oC where the polymer is able to flow into the intergranular space.

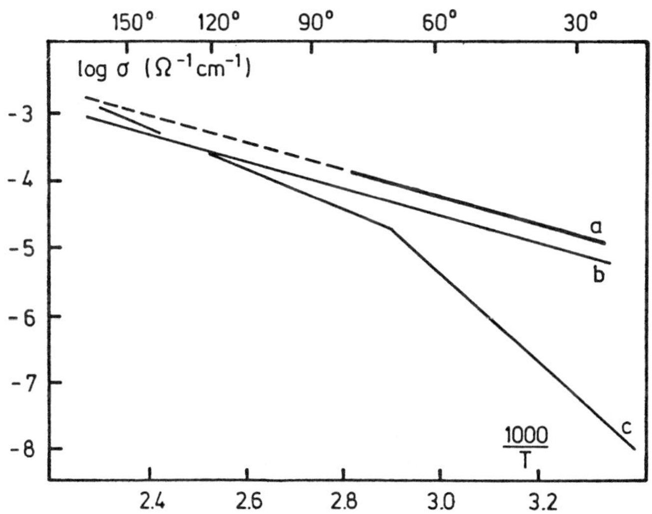

Figure 5. Ionic conductivity of mixed phase electrolytes compared with that of the pure polymer. a) Li_2S-B_2S_3-LiI glass + 20 v/o polymer, b) Li_3N + 4 v/o polymer, c) pure $LiCF_3SO_3(PEO)_{10}$.

Even with as little as 4 volume percent polymer, it is possible to make a stable pellet (pure Li_3N would require temperatures above ~600 ^oC - this would also increase the danger of contamination greatly). Both the mixed phase electrolyte based on Li_3N and that based on Li_2S-B_2S-LiI glass show a room temperature conductivity (σ_{25}~10^{-5} $\Omega^{-1}cm^{-1}$) higher than that of the polymer by a factor of about 300. The conductivity has no break at 70 ^oC but is characterised by a single activation energy (0.42 eV) indicating that the polymer chain motion is no longer a limiting factor. The activation energy is considerably below that of the polymer at high temperatures (0.62 eV) and is even further below the 1.25 eV characterising the polymer at low temperatures. This demonstrates that some of the theoretical advantages of mixed phase electrolytes can indeed be realised. The conductivity is, however, still about 400 times smaller than that of pure Li_3N at room temperature and the activation energy is higher than that in bulk Li_3N (0.26 eV). Possibly an interfacial process is important to the conduction as well as the bulk conduction in the Li_3N grains. When using Li_3N with grainsizes of only a few μm in stead of ~50 μm, the Li^+ mobility decreases (σ_{25}~10^{-6} $\Omega^{-1}cm^{-1}$, E_a~0.50 eV) because the interfacial transport becomes relatively more important.

The same conclusions can be drawn from the glass/polymer combination (E_a = 0.43 eV).

The results show that it is possible to a certain degree to realize the proposed advantages of mixed phase electrolytes in practical experiments. If this type of electrolyte is going to be used technologically, the problems with the still not satisfactory interface will have to be solved. Another major practical problem is that the mixed phase electrolytes with very low polymer contents are not as easy to process into larger sheets as the pure polymer because of the high content of inorganic powder.

Reference:

01. Steen Skaarup, J. Mater. Educ. 6, 667 (1984).

02. B.B. Owens, J.E. Oxley and A.F. Sammels, Topics in Appl. Phys. 21, 67 (1977).

03. T. Lapp, S. Skaarup and A. Hopper, Solid State Ionics 11, 97 (1983).

04. F.W. Poulsen, N. Hessel Andersen, B. Kindl and J. Schoonman, Solid State Ionics 9 & 10, 119 (1983).

05. H. Wada, M. Mentrier, A. Levasseur and P. Hagenmuller, Mat. Res. Bull. 18, 189 (1983).

06. J.E. Weston and B.C.H. Steele, Solid State ionics 2, 347, (1981).

07. M.O'Keefe and B.G. Hyde, Phil. Mag. 33, 219, (1976).

08. U. von alpen and M.F. Bell, Solid State Ionics, 3 & 4, 259, (1981).

09. J.L. Briant and G.C. Farrington, J. Electrochem. Soc. 128, 1830 (1981).

10. J.R. Walker and C.R.A. Catlow, Phil Mag. A43, 265 (1981).

11. F.W. Poulsen, in, Transport-Structure Relations in Fast Ion Mixed Conductors, Proc. of the 6th Risø International Symposium on Metallurgy and Material Science 67 (1985).

12. M. Armand, Solid State Ionics 9 & 10, 745 (1983).

13. T.R. Jow and C.C. Liang, Solid State Ionics, 9 & 10 695 (1983).

14. D. Fauteux, Solid State Ionics, 17, 133 (1985).

15. B. Knutz and S. Skaarup, Solid State Ionics 18 & 19 783, (1986).

SOLID STATE BATTERIES

Steen Skaarup
Physics Laboratory III
Technical University of Denmark
DK-2800, Lyngby, Denmark

Abstract: **The conventional, commercially available batteries mostly use the solid/liquid/solid design for the anode/electrolyte/cathode. The discovery of solid state ionic conductors for use and electrolytes has made other types of designs possible, for instance the reverse design: liquid/solid/liquid used in the construction of the sodium/sulphur battery. The existence of cathode compounds with some ionic mobility in the solid state has made even all solid state designs possible.**

Among the potential advantages of all solid state designs are: mechanical stability, no gassing or leakage, reduced possibility of parasitic side reactions f simple cell designs make high energy densities possible. The disadvantages include the difficulty of achieving good contact between two solid phases and the fact that operation at higher than room temperature may be required.

The emphasis is on alkali metal systems for secondary batteries using intercalation compounds as positive electrodes. The process limiting battery performance under different conditions are discussed as well as theoretical and experimental results on composite cathodes consisting of solid electrolyte mixed with intercalation compound.

Introduction:

Conventional battery designs usually employ and liquid electrolyte (the commercially most imp battery system is the lead-acid system: Pb(s)/H_2 whereas the dominating primary battery system is the Leclanche cell: Zn(s)/NH_4Cl(aq) + $ZnCl_2$(aq)/MnO_2(s)).

One of the reasons for using liquid electrolyte is that aqueous ionic solutions have high ionic conductivities. The discovery of solid state fast ionic conductors with appreciable conductivity in the crystalline state has made novel electrochemical cell designs without liquid electrolytes possible. One example is the sodium/sulphur battery which uses liquid electrodes and solid electrolyte: Na(l)/$Na_2Al_{22}O_{34}$(s)/S(l) - thereby reversing the conventional design. The cell employs the solid state Na^+ ion conductor sodium beta aluminia as separator and electrolyte.

By using intercalation (insertion) compounds for the positive electrode it is possible to go one step further and construct all solid state cells with no liquid components. An intercalation compound is characterized by the ability of the host lattice to receive foreign atoms or molecules without altering the topology of the host structure. The insertion of foreign species is usually accompanied by a volume expansion of the order of 10%.

Figure 1 shows the structure of the intercalation compound titanium disulphide, TiS_2. The sulphur atoms form a hexagonally close packed structure. Every other plane between sulphur layers contains octahedrally coordinated titanium. The surplus sites in the van der Waals gap can accomodate a wide range of molecules and ions.

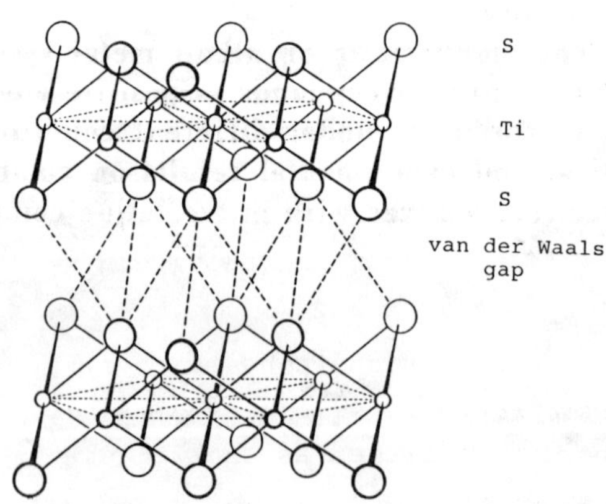

Figure 1. The hexagonal structure of TiS_2.

For battery purposes, the most important guest species is the Li^+ ion - the TiS_2 lattice can receive one lithium atom for each titanium without any breaking of strong bonds or changes in structure. The Li^+ ions are able to move in the lattice although the diffusion coefficients are 4-5 orders of magnitude smaller than in the best fast ion conductors.

Titanium disulphide is used as example because it has an exceptionally simple structure, but there are many other compounds capable of inserting Li^+ ions, i.e. a number of vanadium oxides |1|.

Potential Advantages of The All Solid State Design:

There are a number of theoretical advantages to using solid state batteries with intercalation compounds:

- Problems with leakage and gassing are avoided.
- Solid state reactions are usually much more specific than reactions in aqueous or organic solutions. This means fewer possible parasitic side reactions leading to self-discharge and irreversible loss of capacity.
- Solid electrolytes which are thermodynamically stable towards alkali metals can be found. This points to the possibility of constructing cells with low weight and high voltage.
- The intercalation cell reaction is the simplest possible: $xLi + TiS_2 \rightleftharpoons Li_xTiS_2$ $0 \leq x \leq 1$, where the degree of intercalation, x, can vary continuously between the limits. Since no new compounds or even phases are formed, a high degree of reversibility may be attainable.
- The electrolyte does not participate in the cell reaction as in the case of sulphuric acid in the lead-acid system. This may save weight.
- In true solid electrolytes, only one ion is mobile. The lack of anion mobility prevents the buildup of concentration gradients in the electrolyte, thereby avoiding limiting current effects.
- The all solid construction may result in a simple and mechanically stable battery design since the solid electrolyte acts also as separator and since each cell does not need a separate container. A simple cell design may also result in higher practical energy densities as the weight of the nonactive components can be minimized.

- The thermal stability at higher temperatures is superior to aqueous or organic electrolyte systems.

There are, however, also disadvantages to using solid state batteries:

- Achieving and retaining reliable interfacial contact between two solid phases is much more difficult than if one of the phases is liquid.
- The generally smaller ionic mobility in the solid phase may necessitate operating at elevated temperatures (~100°C) or the use of extremely thin layers.

Energy Content and Ionic Transport in the Solid State Battery

Figure 2 shows a schematic concept of a solid state battery. The $Li/Li_3N/TiS_2$ system is used as example. The structure is the simplest possible with a thin membrane of solid ionic conductor functioning both as electrolyte and as separator.

The amount of energy per unit weight of the battery system is proportional to the average voltage during a discharge, to the utilization of the active materials and inversely proportional to the mass of the components.

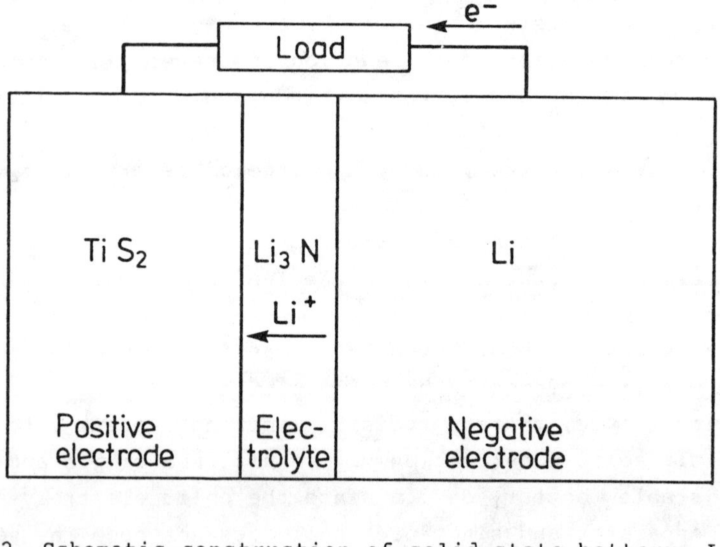

Figure 2. Schematic construction of solid state battery. Lithium is the anode and titanium disulphide the cathode during discharge. The current directions shown correspond to the discharge process.

The theoretical specific energy for the Li/TiS_2 system is determined by measuring the electromotive force as a function of the degree of Li^+ intercalation (figure 3,|2|) and using the mass of stoichiometric amounts of lithium and titanium disulphide (the solid electrolyte does not participate in the cell reaction). The specific energy calculated in this way is 480 Wh/kg. The corresponding value for V_6O_{13} which can insert 8 lithium atoms reversibly is 890 Wh/kg. These values should be compared to 170 Wh/kg for the lead-acid system.

In order to be of practical use, the battery must be able to be discharged with currents of a reasonable size so that the energy can be delivered to a load at a rate suitable for the application. This depends on sufficiently rapid transport of electrons and ions in the system. The electronic transport in the lithium anode will always be fast enough unless extremely thin layers are used - this is also the case when using a compact titanium disulphide cathode. Other intercalation compounds may require the addition of some form of conducting graphite. The transport of Li^+ ions is much less rapid and is one of the main considerations when designing batteries for practical use.

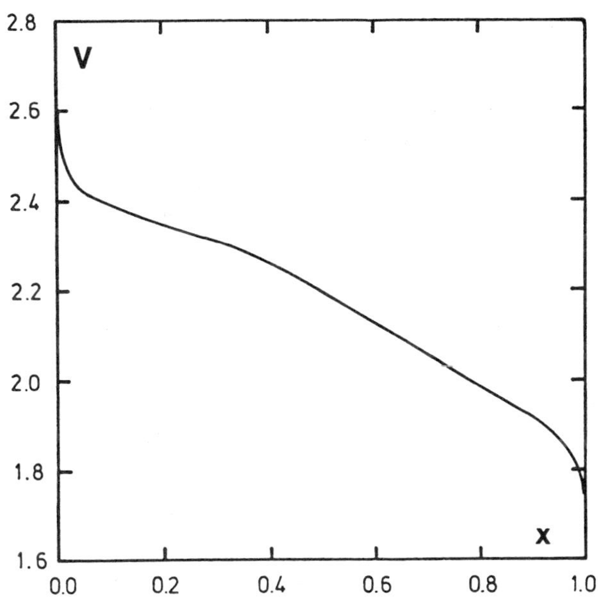

Figure 3. Open circuit voltage with respect to metallic lithium of Li_xTiS_2 as a function of the degree of Li^+ intercalation.

During a discharge process, a Li^+ ion will experience 5 successive barriers to motion:

- Transport in the bulk metal anode. (Lithium).
- Transfer across the anode/electrolyte interface. (Li/Li_3N).
- Transport in the separator electrolyte. (Li_3N).
- Transfer across the electrolyte/cathode interface. (Li_3N/TiS_2)
- Transport into the cathode. (TiS_2).

Transport in the Anode:

The transport of lithium atoms in the negative electrode is necessary in order to replenish the voids formed at the interface as Li^+ ions are removed into the electrolyte. If selfdiffusion in the metal is the only transport process, a current density I (A/cm^2) will cause depletion of the metal after the time t_o (s) given by the relation |3|:

$$I\sqrt{t_o} = \frac{F\sqrt{\pi}}{2} C_o\sqrt{D}$$

where F is Faradays constant, C_o the concentration of metal atoms in the anode (C_o = 0.0764 mol/cm^3 for Li) and D is the selfdiffusion coefficient ($D = 5.6*10^{-11}$ cm^2/s at room temperature for Li.) Using the lithium values at room temperature and assuming a discharge time of 10^4 seconds. The limiting current density is 0.49 mA/cm^2. If current densities higher than this (or longer discharge times) are desired it will be necessary to increase the rate of transport - either by applying pressure to the system or by raising the temperature. The pressure required at room temperature and I = 1 mA/cm^2 is of the order of 500 kg/cm^2. This may be a severe limitation. At temperatures above about 100 ^{o}C, springs incorporated in the cell can easily maintain contact at current densities up to at least 5 mA/cm^2 because the metal becomes softer.

An alternative to metallic lithium is to use alloys such as LiAl or soft, low-melting systems based on Woods metal. This increases the weight of the anode considerably, however, thereby losing one of the advantages of using lithium.

Transfer Across the Anode/Electrolyte Interface:

Since Li_3N is stable in contact with metallic lithium there is no tendency to form a reaction layer that might block the transfer of ions. The Li/Li_3N interface therefore functions well with little transfer resistance and is able to be cycled hundreds of times without any deterioration.

If the electrolyte used is not thermodynamically stable towards lithium, there will be a tendency to form a reaction layer of very stable lithium compounds (e.g. Li_2O, LiOH, Li_2CO_3, LiCl or LiF) at the interface, thereby increasing the internal resistance of the cell. This may be the case when using ionically conducting polymers such as poly ethylene oxide containing lithium salts ($LiCF_3SO_3$, $LiClO_4$ etc.). If the reaction layer becomes progressively thicker, it will eventually block the ionic transport. Since many lithium salts show some ionic conductivity, a layer of constant, low thickness may not be detrimental to the operation of the cell. Battery systems containing lithium in contact with liquid electrolytes, such as $LiClO_4$ in propylene carbonate or thionyl chloride in the high energy density primary battery system $Li/SOCl_2$, are stable only because of the formation of thin protective layers of for instance LiCl.

Ionic Transport in the Separation Electrolyte:

In a true crystalline fast ion conductor the concentration of ions is very large (Li^+ in Li_3N: 112 moles/liter) and does not change during the passage of an ionic current. Since only the cations move there is no formation of concentration gradients leading to depletion of ions at a limiting current value. The main consideration in the design is to make the electrolyte so thin that it does not contribute appreciably to the ionic resistance and to the weight of the battery - this must, however, be balanced by considerations of mechanical stability. For a current density of 1 mA/cm^2 and a specific ionic conductivity of $3*10^{-4}\ \Omega^{-1}cm^{-1}$ (Li_3N at 25 ^oC), a voltage drop of 100 mV, which may be acceptable, corresponds to a thickness of 0.3 mm which is a technologically reasonable value.

If the separator electrolyte consists of a polymer ionic conductor with appreciable anion motion, concentration gradients causing electrolyte depletion will define a limiting current density (|4|) which is inversely proportional to the thickness of the electrolyte, d. It also depends on the concentration of ions, C, on the salt-diffusion coefficient, D, and on the anion transport number t_2:

$$I = \frac{2FCD}{t_2 d}$$

This limits the use of the present polymer electrolytes to a thickness of 50-100 μm even at elevated temperatures. This limitation is partially offset by the fact that polymer electrolytes of this thickness are easy to fabricate. The resistance to dendrite formation of the thin polymer layers is enhanced by the phenomenon "soft dendrites": Temporary, thin metal dendrites that disappear again.

Transfer Across the Cathode/Electrolyte Interface:

The interface between two crystalline compounds, e.g. Li_3N and TiS_2 is difficult to establish so that it will function satisfactorily for many battery cycles. A great improvement in performance is obtained by using a thin layer of ionically conducting polymer between the two phases. This ensures contact over hundreds of charge/discharge cycles because of the flexibility of the polymer at temperatures over ~70 ^oC.

Ionic Transport in the Intercalation Cathode:

Diffusion processes in general are characterized by time constants $t = d^2/D$ where d is a typical diffusion length, such as the thickness of an electrode or the diameter of a particle, and D is an effective chemical diffusion coefficient.

When an intercalation electrode is discharged, an approximately parabolic concentration profile of diffusing Li^+ ions is formed |5|. The profile as a whole moves towards gradually higher values as the discharge proceeds (figure 4).

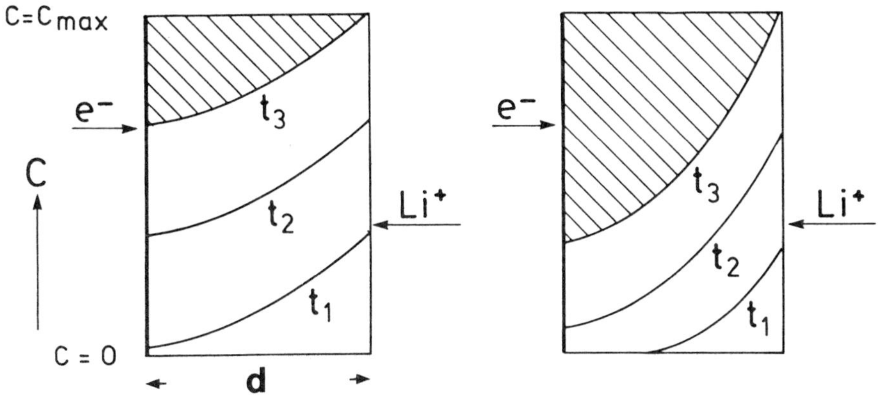

Figure 4. Concentration profiles of Li^+ ions in an intercalation electrode for succesive discharge times t_1, t_2 and t_3. At t_3, the concentration corresponds to the maximum degree of intercalation, and the discharge stops. The hatched areas represent the capacity lost. The left discharge current is half of that to the right.

When the degree of intercalation at the interface between separator and electrolyte corresponds to the maximum concentration, C_{max} ($\sim Li_{1.0}TiS_2$), the electrode can no longer receive lithium ions and the voltage drops as the discharge stops. This leaves part of the inner electrode still not fully intercalated corresponding to a utilization less than 100%. The amount of capacity lost depends on the load factor, L, which is proportional to the current. L is equal to t/T_s, where t is the time constant of the electrode and T_s the stoichiometric discharge time that would be required to discharge all the intercalation compound in the electrode at the given current.

The utilization, U, depends on the load factor as depicted in figure 5. In the region of most interest, where more than 50% of the active material is utilized, the simple formula:

$$U = 1 - L/3$$

is approximately valid. At higher currents, the utilization is equal to $\pi/4L$.

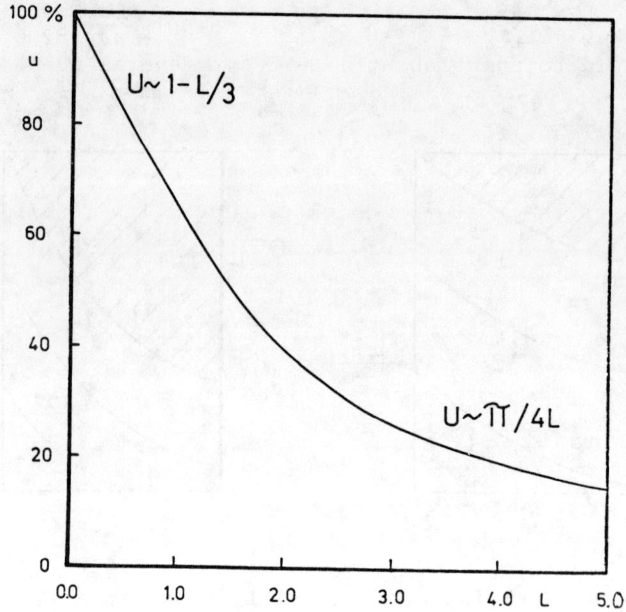

Figure 5. Electrode utilization as a function of the load factor for an intercalation electrode.

The values for the Li^+ diffusion coefficient in TiS_2 are shown in figure 6 (|6|). Since D_{Li} is only about 10^{-10} cm^2/s, a discharge time of about 3 hours corresponds to the unreasonably low optimal electrode thickness of only 10 μm at 25 °C.

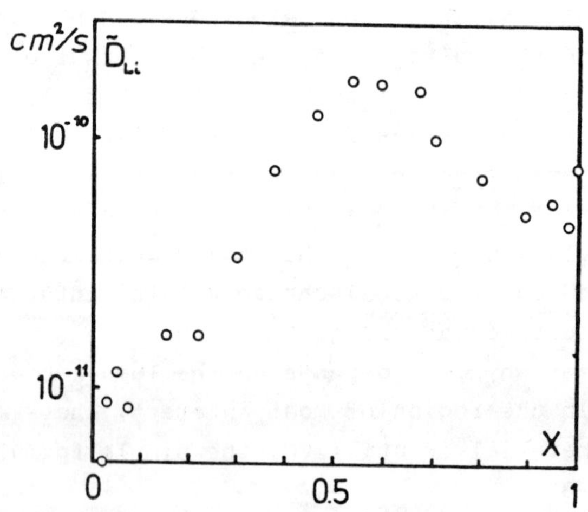

Figure 6. D_{Li} in Li_xTiS_2 as a function of x (25 °C).

D_{Li} can be increased by raising the temperature, and laboratory solid state cells with compact 30-50 μm TiS_2 electrodes have been able to function at 170 ^{o}C for several hundred cycles |7|.

A better way of improving ionic transport in the cathode is to incorporate a certain amount of liquid or solid electrolyte which increases the rate of ionic transport. The resulting structure is called a porous or composite electrode respectively. Porous electrodes are used in practically all commercial batteries.

The ideal composite electrode consists of two contiguous, interconnecting networks (figure 7). The ions are transported mainly in the electrolyte phase - the electrons in the intercalation phase, and the cell reaction takes place along the large common interfacial area. Under simplifying conditions, it can be shown that the transport is characterized by an effective Li^+ coefficient of diffusion given by |8|:

$$D_c = \frac{kS}{FvC_{max}}$$

where k is the slope of the emf-curve with respect to the degree of intercalation (figure 3), S is the ionic conductivity of the electrode as a whole, F is Faradays constant, v is the volume fraction of intercalation compound and C_{max} the saturation concentration of the inserted ion.

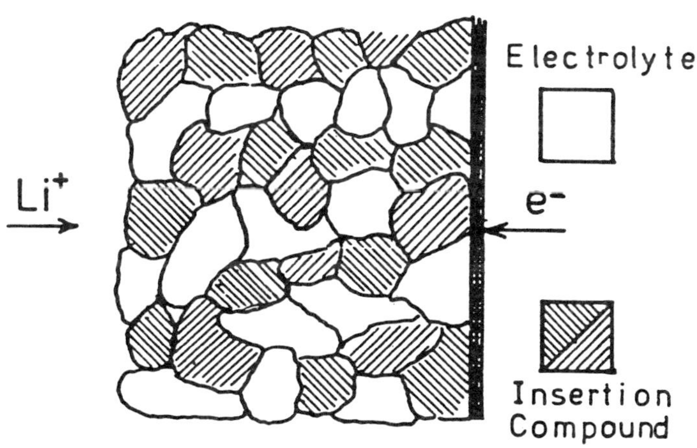

Figure 7. Schematic structure of solid state composite electrode.

All the parameters except S are known from bulk properties or can be controlled. The ionic conductivity of the electrode is equal to that of the pure crystalline electrolyte multiplied by a factor depending on the tortuosity of the conduction path and the volume fraction of electrolyte. (The resistance from the transfer across the interface is assumed to be negligible). The values of the correction factor calculated by using classical theories of conduction in heterogeneous media and from different percolation theories vary widely |9|. A reasonable estimate is 0.2-0.3. This corresponds to an effective diffusion coefficient of $1\text{-}4 \cdot 10^{-6}$ at 140 °C for a composite electrode of 50/50 volume percent Li_3N and TiS_2. This would imply that the possible cathode thickness could be increased from 10 μm to 1-2 mm making handling and fabrication much easier.

Experimental Results:

Figure 8 shows the results of cycling a $Li/Li_3N/Li_3N+TiS_2$ solid state laboratory cell at 140 °C. For the first 300 cycles the cyclability is very good - corresponding to a cycling efficiency of 99.87%. This is a reasonable value for practical applications.

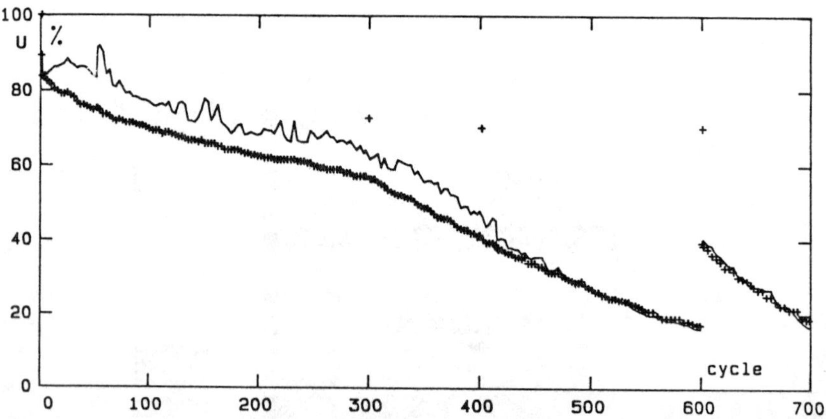

Figure 8. Charge (unbroken line) and discharge capacities (+) for a $Li/Li_3N/TiS_2+Li_3N$ solid state battery as a function of cycle number. The current was lowered after 600 cycles.

The charge capacity is equal to the discharge capacity when impurities are carefully excluded (cycles 500-700). This shows that one of the advantages of solid state cells - the smaller likelihood of interfering side reactions - can be realized in practice.

The measured effective Li^+ diffusion coefficient is about a factor of 10 lower than the theoretical estimates, indicating that the electrode/electrolyte interface contact functions less than optimally. Figure 9 shows the effective diffusion coefficient as a function of temperature at different cycle numbers. After 700 cycles the diffusion coefficient has decreased by a factor of about 10 and the activation energy has increased from 0.39 to 0.80 eV. In an ideal composite electrode, the activation energy would be the same as in the electrolyte phase (E_a in Li_3N: 0.26 eV). The result shows that the contact between the components of the composite electrode has deteriorated strongly - perhaps because of the expansion and contraction of the intercalation compound as it is cycled. This illustrates one of the main disadvantages of the solid state concept.

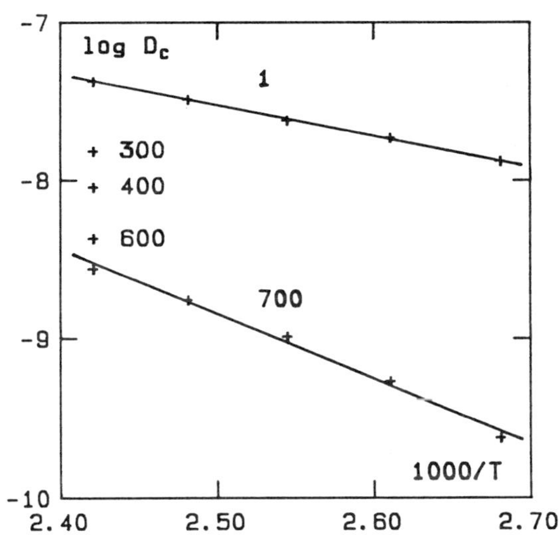

Figure 9. Effective Li^+ diffusion coefficient (cm^2/s) for a composite electrode as a function of cycle number (marked on figure) and temperature.

Similar results have been obtained using polymer electrolyte instead of crystalline electrolyte in the cathode |10|. This limits the thickness to about 0.1 mm but the mechanical properties of the polymer makes the fabrication of sheets of this thickness possible. The very thin cells still have the problem that many cells have to be stacked to form a large battery. This means the use of many layers of very thin lithium and current collector metal if the specific energy is to retain its high value.

At present there are no secondary solid state batteries being manufactured commercially. The Li/V_6O_{13} system using polymer electrolyte, which has a theoretical energy content more than 5 times that of the lead-acid system, is being developed towards industrial production by several manufacturers |11|. The designs aim at practical specific energies of 100-250 Wh/kg (lead-acid: 25-40). Reliable batteries with these characteristics could be used for a number of purposes and might also make the electric car a feasible alternative.

Acknowlegements:

Clarifying discussions with my coworkers at the Institute of Physical Chemistry are gratefully acknowledged.

References:

01. K. West, B. Zachau-Christiansen, M.J. Landeira and T. Jacobson, *in*, Transport-Structure Relations in Fast Ion and Mixed Conductors. *Proceedings of the 6th Risø International Symposium on Metallurgy and Materials Science*, (1985) 265.

02. T. Jacobson, K. West, and S. Atlung, *Electrochemica Acta* 27, 1007, (1982).

03. T. R. Jow, and C.C. Liang, *J. Electrochem. Soc.* 130, 737, (1983).

04. S. Atlung, in, *Solid State Batteries* ed. C.A.C. Seqira and A. Hooper, 129, (1985).

05. S. Atlung, K. West and T. Jacobsen, *J. Electrochem. Soc*. 128, 1311 (1979).

06. K. West, S. Atlung, and T. Jacobson, *J. Electrochem. Soc.* 126, 2875 (1982).

07. B. Knutz and S. Skaarup, *Solid State Ionics* 9 & 10 371, (1983).

08. S. Atlung, B. Zachau-Christiansen, K. West, and T. Jacobsen, *J. Electrochem. Soc.* 131, 1200 (1984).

09. Boye Knutz and Steen Skaarup, Solid State Ionics 18 & 19, 783, (1986).

10. K. West, B. Zachau-Christiansen, T. Jacobsen and S. Atlung, *J. Power Sources*, 14, 235, (1985).

11. M. Gauthier, A. Belanger, D. Fauteux, M. Duval, B. Kapfer, M. Robitaille, R. Bellemare and Y. Giguere, Extended Abstracts, *Third International Meeting on Lithium Batteries*, Kyoto, 238, (1986).

PART II

SHORT PRESENTATIONS

THE REFRACTIVE INDEX AND OPTICAL GAP OF ANNEALED AMORPHOUS SILICON

R. Salam, S. A. Rahman, M. R. Muhamad, A. R. Danker
Solid State Laboratory, Physics Department
University of Malaya, 49100
Kuala Lumpur, Malaysia

Abstract: **Amorphous silicon (*a*-Si) prepared by d.c. glow discharge decomposition gas is thermally annealed up to 600#C. The index of refraction n, as measured by the Brewsters angles technique is found to have a minimum (n = 2.05) at an annealing temperature Ta 210#C. Corresponding values of the optical gap E_{opt} Ta. Results are compared with other workers and a mechanism to explain this phenomenon is proposed.**

n-Cu_2O/p-Cu_x S HETEROJUNCTIONS THIN FILM SOLAR CELL

W. Siripala and K. Premasiri Kumara
Department of Physics
University of Kelaniya
Kelaniya, Sri Lanka

Abstract: **Electrodeposited cuprous oxide is an n-type semiconducting material in contrast to the well established result of p-type conducting thermally grown cuprous oxide. n-Cu_2O/p-Cu_xS heterojunction thin film solar cell was fabricated and its photovoltaic properties were investigated. We have obtained the values of V_{oc} = 100 mV and I_{sc} = 150 lA/cm^2 under AM 1 illumination. The performance of the cell is very sensitive to heat treatment.**

Introduction:

Experimental and theoretical studies of Cu_2O solar cells have been reported previously[1-3], however, the experimentally obtained efficiencies were much less than the theoretically predicted values. All the previously reported works were based on the thermally grown p-type Cu_2O only. Nevertheless, it has been reported recently that electrodeposited Cuprous Oxide films produce n-type conductivity and the heterojunction n-Cu_2O/p-Cu_xS can be easily fabricated.[4,5] In this investigation we have carried out several experiments in order to examine the several parameters those influencing the efficiency of the n-Cu_2O/ p-Cu_x S thin film solar cell. We have observed that the cell performance can be improved significantly by the heat treatments and the surface pre-treatments.

Experimental:

Cuprous Oxide thin films were deposited on Cu substrates using the previously described technique.[4] Copper plates were initially polished with sand paper and then washed, degreased and rinsed with distilled water. The electrodes were then dipped in an electrolyte containing 0.1 M $CuSo_4$ with added few drops of NaOH for about 8 hours. After the film deposition, the electrodes were rinsed in distilled water and then heated in air at 50°C for few minutes. Thin Copper Sulphide (Cu_xS) coatings were deposited on the Cu_2O films by spraying an Ammonium Sulfide solution over the surface of Cu_2O films. Electrical contact to the front surface was made by using a mechanically pressed gold plated Copper grid, while the back contact was made to the Copper substrate. During the heat treatments the front contact was removed. We have observed that the chemical etching of the Cu_2O sample with KCNS considerably improves the efficiency of the cell. The Cu_2O sample were dipped in 1M KCNS solution for 1 second and then rinsed in distilled water for a long time. Afterwards the sample was heated in air and the Cu_xS coating was made.

Sheet resistance measurements during the heat treatments were made using the conventional four-probe technique. The probes were made of gold plated Copper wires and it was also ensured that the I-V characteristics were linear over the ranged used in this experiment.

Results and Discussion:

Fig.1 shows the light and dark current-voltage characteristics of a Cu_2O/Cu_xS thin film solar cell. Intensity of illumination is 100 mW/cm^2. The large dark current indicates that the junction is 'leaky' and hence the performance of the cell is seriously limited. The I-V behaviour was different from sample to sample though the basic characteristic remained to be the same. The current transport mechanisms may be quiet complicated and obtaining quantitative results are doubtful. However, the general 'diode-like' character was demonstrated.

Fig.2 shows the variation of the open-circuit voltage (V_{oc}), the short-circuit current (I_{sc}) and the sheet resistance with the period of heat treatment at temperature 50°C. The sheet resistance values clearly show that the shunting resistance of the interface is modified by the heat treatment and correspondingly V_{oc} and I_{sc} are being changed. A low open-circuit voltage is associated with the low shunting resistance. The major contribution to the low shunting resistance might come from the interface states. This is supported by our observation that chemical etching improves the performance of the cell significantly. The chemical etching may reduce the density of interface states at the junction. This would result in the decrease in the recombination losses and increase in I_{sc}. Similarly heat treatments may reduce the interface states. We have observed that the high temperature heat treatments give better results than the low temperature treatments. However, the exposure of the cell for a longer time would dgrade the cell if the temperature is above 150°C.

One other possibility for the low shunting resistance might be the effect of grains which might produce conducting channels that extends deep into Cu_2O and hence resulting the low shunting resistance.

Conclusion:

In conclusion, we have demonstrated that the efficiency of the n-Cu_2/p-Cu_xS thin film solar cell can be improved significantly by proper heat treatments and surface pre-treatments. The semiconducting materials involved in this cell is less expensive compared to many other solar cell semiconducting materials and the fabrication procedure of the cell is also very simple. Therefore this system is interesting for further studies.

Acknowledgement:

This work was supported by a research grant from the Natural Resources, Energy and Science Authority of Sri Lanka.

References:

1. L.C. Oslen, F.W. Addis and W.Miller; Solar cells 7, 247 (1982).
2. M. Fujinaka and A.A. Berezin; J. Appl. Phys. 54, 3582 (1983).
3. L.C. Olsen, R.C. Bohara and M.W. Uriei; Appl. Phys. Lett. 34, 47, (1979).
4. W. Siripala and J.R.P. Jayakody; Solar Energy Materials 14, 23 (1986).
5. W. Siripala; Solid State Communications, (in press).
6. A.L. Faherenbruch and R.H. Bube; J. Appl. Phys. 45, 1264 (1974)

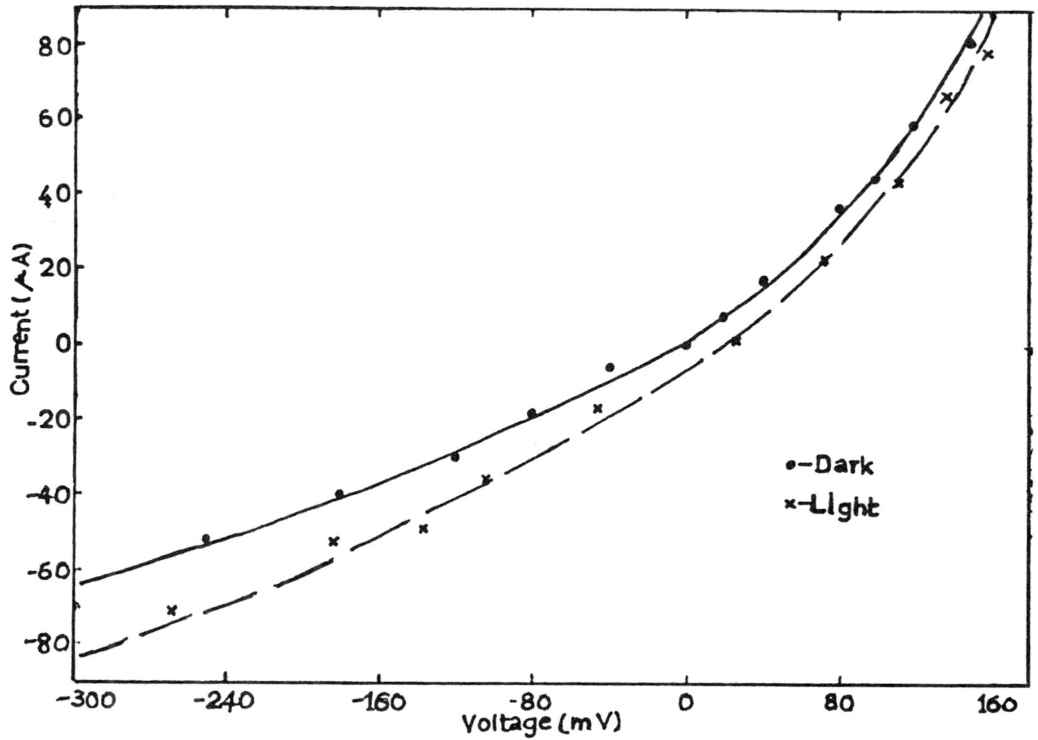

Fig: 1 Light and Dark Current-Voltage Characteristics of n-Cu_2O/P-Cu_xS heterojunction

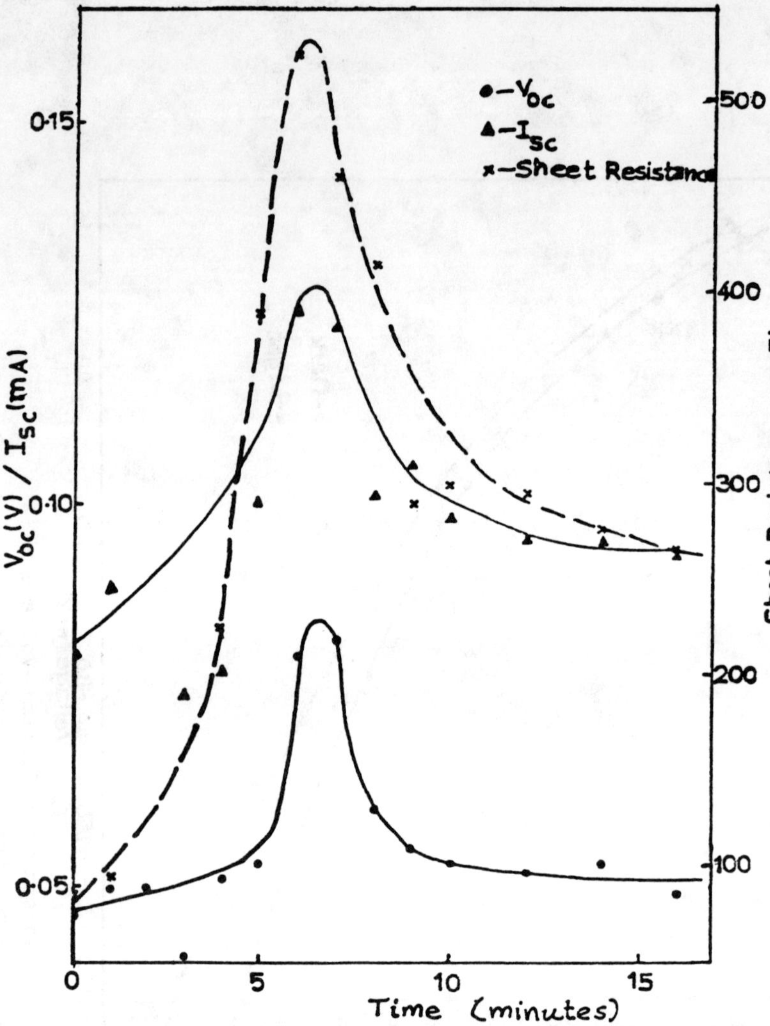

Fig: 2 Variations of V_{oc}, I_{sc} and sheet resistance of Cu_2O/Cu_xS solar cell with the period of heat treatment at 50 °C .

PHOTOCATALYTIC NITROGEN REDUCTION WITH VISIBLE LIGHT ON HYDROUS FERRIC OXIDE

S. Punchihewa
Institute of Fundamental Studies
Hantana Road
Kandy, Sri Lanka

Abstract: **An aqueous suspension of hydrous ferric oxide is found to photocatalyze reduction on molecular nitrogen to ammonia with the concommitant oxidation of water under visible light. The yield of ammonia reaches a maximum after ca 1h irradiation, and decreases thereafter. The optimum pH for NH_3 production is around 10. In addition to ammonia, nitrate is also formed in this reaction, which increases with time. The nitrate formed shifts the flat band potential of the catalyst towards the less negative region, which is unfavorable for N_2 reduction. The catalytic activity of this material is attributed to strongly negative flat band potential and chemisorption of N_2.**

Introduction:

In recent years, considerable efforts have been made on the development of systems which are capable of collecting and storing solar energy in the form of chemical potential. Among these systems, the photocatalytic synthesis of ammonia from water and nitrogen has received some attention. It has been reported that iron doping gives the highest yields of NH_3 on TiO_2[1-4] while a mixed Oxide catalyst $NiO\text{-}SrTiO_3\text{-}RuO_2$ is reported to have even a higher activity[5]. In all these reports, the yield of NH_3 obtained is very small. Further it has been shown that the catalytic activity rapidly decreases with time.

In this presentation, the first observations of a system where N_2 reduction occurs with visible light is discussed. An aqueous suspension of hydrous ferric oxide is found to photoreduce N_2 to NH_3 with concommitant oxidation of water. In addition to ammonia, nitrate is also formed during this reaction.

Experimental:

Hydrous Ferric Oxide was prepared by the addition of 0.1 M NaOH to a solution of 0.1M $FeCl_3$. (A.R. chemicals were used wherever possible). The precipitate was washed several times with water and used for irradiation. Photocatalytic reactions were performed in 100 ml round bottom flasks containing the freshly prepared catalyst suspended in double distilled water. The pH of the medium was adjusted to Ca 10 by the addition of NaOH.

The reaction medium was vigorously - stirred throughout the reaction. Irradiation was performed with a 100W Tungsten filament lamp under a slow stream of purified N_2 gas. The gas purified by passing it thorough a series of wash bottles containing chromic acid, KOH, HCl and distilled water. After the irradiation 2ml of 10% NaOH was injected into the flask and the NH_3 formed was distilled off into a trap containing 10ml of 0.1M HCl solution. The yield of NH_3 was determined - colorimetrically by the

Indophenol Blue method. Total Nitrate and Nitrite content was also determined colorimetrically by the Cadmium Reductor method. The variation of NH_3^- formation as a function of time and pH was investigated.

Results and Discussion:

The results for the reduction of molecular nitrogen to ammonia are shown in Fig.1, and Fig.2 shows the variation of ammonia formed as a function of pH. The reactions involved are:

$$N_2(g) + 6H^+(aq) + 6\bar{e} \longrightarrow 2NH_3(aq) \text{ —— } (1)$$

$$3H_2O(l) + 6\overset{+}{h} \longrightarrow 3/2O_2 + 6H^+ \text{ —— } (2)$$

The rates of NH_3 formation were initially high, but decreased rapidly. The gradual decrease of the yield is probably due to the photocatalytic decomposition of NH_3. It was found that samples irradiated for longer periods of time, contain detectable amounts of nitrates. This is probably due to the oxidation of ammonia. The photoreduction of nitrogen was possible even with air as the source of N_2. It was also found that trace quantities of NO_3^- in the solution shifts the flat band pontential of $Fe_2O_3(H_2O)_n$ towards the less negative direction (Fig.3). The highly negative flat band is obtained only under strongly alkaline conditions, which enhances the N_2 reduction reaction. We found that the catalytic activity decreased with drying and heating of the catalyst. It seems to depend on the high degree of hydration of the colloidal form of the catalyst, and chemisorption of N_2 on the catalyst seems to be an important step in this reaction.

References:

1. S. Grayer and M. Halmann, J. Electronal Chem.; 170, 363, (1984).
2. G.N. Schrauzer and T.D. Guth, J. Amer. Chem. Soc., 99, 7189 (1977).
3. H. Miyama, N. Fujii and Y. Nagal, Chem. Phys. Lett., 74, 523 (1980).
4. G.N. Schrauzer, N. Strampach, L.N. Hui, M.R. Palmer and J. Salehi, Proc. Natl. Acad. Sci, USA, 50, 3873, (1983).
5. Q.S. Li, K. Domen, S. Natio, T. Onishi and K Tamaru, Chem. Letts., 321, (1983).

VARIATION OF NH_3/NO_3^- FORMED AS A FUNCTION OF TIME

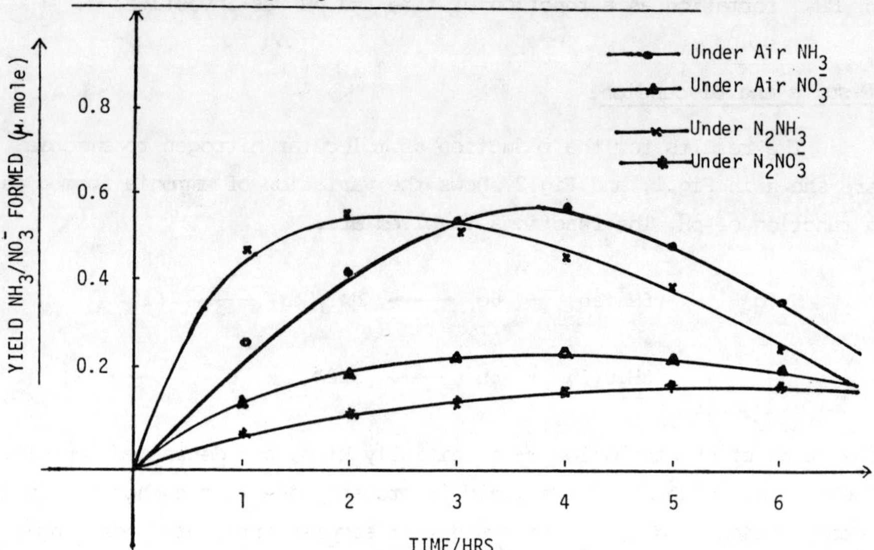

Fig. 1

VARIATION OF AMMONIA FORMED AS A FUNCTION OF pH

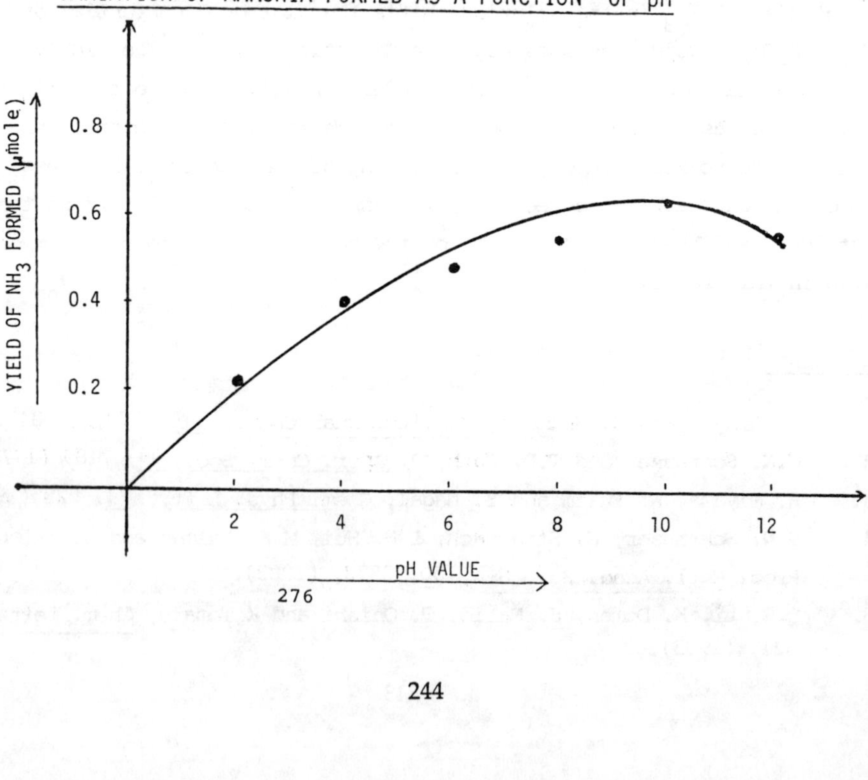

276

THE pH VARIATION OF THE PHOTOCURRENT ON-SET POTENTIAL

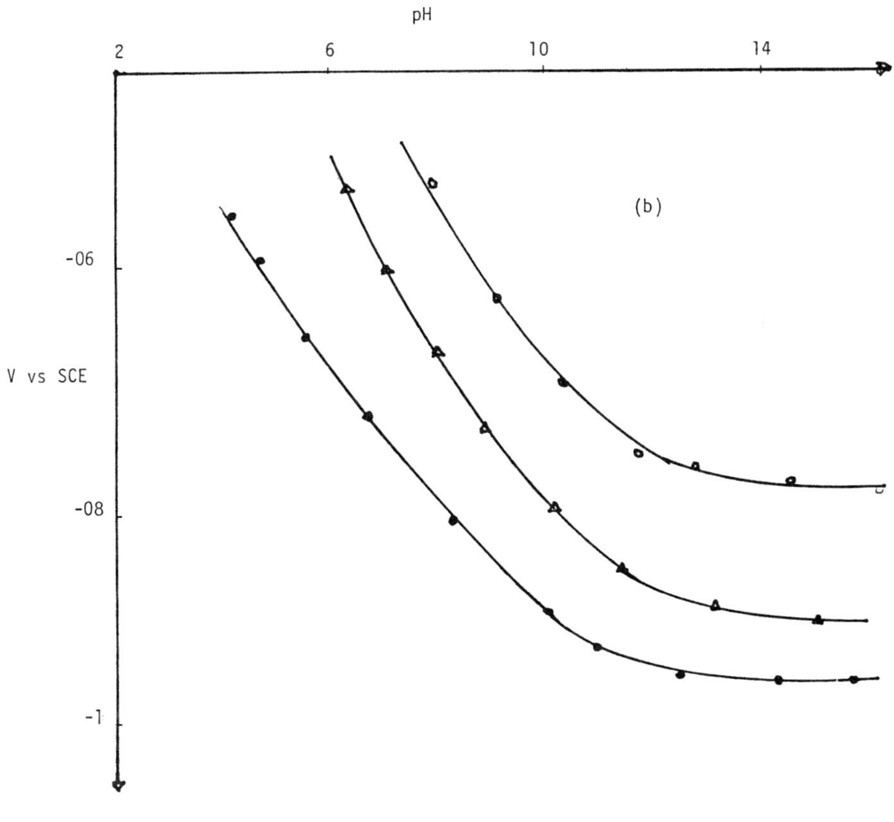

Fig. 3

● - $Fe_2O_3\ (H_2O)_n$

○ - Fe_2O_3

△ - $Fe_2O_3\ (H_2O)_n$ in the presence of NO_3^-

AC CONDUCTIVITY OF $CuInSe_2$

K. Premaratne and S. H. S. P. Samarappuli
Department of Physics, University of Peradeniya
Peradeniya, Sri Lanka

M. A. K. L. Dissanayake
Department of Physics, University of Peradeniya
and
Institute of Fundamental Studies, Kandy, Sri Lanka

Somphong Chatraphorn
Semiconductor Physics Research Laboratory
Chulalongkorn University, Bangkok, Thailand

Abstract: **$CuInSe_2$ samples have been prepared by the horizontal Bridgman method using high purity Cu, In, and Se weighted according to the stoichiometry of $CuInSe_2$. Ohmic contacts were made on the thin disc-shaped samples using Indalloy as well as gold. The d.c. conductivity of the samples measured using four-probe technique was found to be about 0.1 X^{-1} cm^{-1} at room temperature. the a.c. conductivity of both n and p type samples were found to be frequency independent in the frequency range 10 to 100 kHz. the temperature dependence of the conductivity at the d.c. limit indicates that in the temperature range 300 - 500 K hopping conduction at the band edges is predominant whereas in the range 500 - 800 K the conduction appears to be predominantly due to excitation of carriers into the extended states.**

Introduction:

The ternary chalcopyrite semiconductor $CuInSe_2$ has drawn considerable attention in recent years since it shows promise for practical use as a solar cell material [1]. It has a direct energy gap of about 1 eV at room temperature and an absorption coefficient of 10^5 cm^{-1} at photon energies $\gtrsim$ 1.5 eV [2-4]. In recent years a great deal of work has been done on this semiconductor[5]. Various techniques for preparation and characterization of thin films, single crystals, polycrystalline samples of this compound have been reported [5]. There have also been work reported on $CuInSe_2$ based heterojunction solar cells [6]. However not much work has been done to fully understand the conduction mechanisms in this material. Study of the ac conductivity enables one to examine purely bulk dissipative mechanisms since the contact effects become insignificant especially at higher frequencies. Possible electronic dissipative mechanisms that lead to the observed frequency and temperature dependence of the ac conductivity have been reviewed by Mott and Davis [7].

In this work we have studied the frequency dependence of the ac conductivity of $CuInSe_2$ and the temperature dependence of the ac conductivity at the dc limit. An attempt has been made to interpret the observed results on the basis of existing theories on ac conductivity of semiconductors with a view to understand the mechanisms of conduction in this material.

Experimental:

The $CuInSe_2$ samples used in this work were prepared by horizontal Bridgman method. The elements Cu, In and Se of purity 99.999% were weighed according to the stoichiometry of $CuInSe_2$ and sealed in quartz ampoules under 10^{-5} torr vacuum. These were then placed in a tube furnace and the usual Bridgman procedure was followed in order to make $CuInSe_2$ samples. The samples were observed to be polycrystalline with grain size of about 3x3 mm^2. Samples obtained from different runs were either n -type or p -type. The change of conductivity type among different batches could be attributed to the small variations of Se concentration in the samples because Se could easily evaporate even at low temperatures.

The structure of the resulting compound was confirmed using X-ray diffraction. The samples were cut into small discs of about 5x5 mm^2 cross section and 1 mm thickness. These discs were subsequently polished with 5 μ alumina and thoroughly cleaned using standard cleaning techniques.

It was found that evaporated gold gave good ohmic contacts on p-type samples. The n-type samples did not give good ohmic contacts with gold. However, indalloy (44% In, 42% Sn, 14% Cd) was found to give ohmic contacts on these samples.

The resistivity of the samples were measured using the four-probe technique [8]. The experimental arrangement used for the measurement of frequency dependence of ac conductivity is shown in figure 1. It consisted of an HP 4192 A impedence analyzer controlled by an HP 86 B microcomputer, and a Herause tube furnace controlled by an Eurotherm 810 temperature controller.

Results and Discussion:

The resistivity of the samples was about 10 Ω cm at room temperature. The ac conductivity of the samples shown in figure 2 is frequency indepndent in the frequency range 10 - 10^5 Hz. This is observed to be true for all temperatures chosen for this experiment. The existing theories [7] predict that at moderately high temperatures ($>$ 300K) the conduction in amorphous semiconductors is largely due to excitation of carriers into the extended bands and also due to hopping of carriers at the band edges. These theories also indicate that the conduction due to above mechanisms is independent of frequency at least upto 10^7 Hz. The observed results appear to agree with these predictions.

The temperature variation of the conductivity of $CuInSe_2$ at the dc limit is shown in figure 3. The $\log \sigma_{dc}$ vs 1/T graph is observed to have two approximately linear regions with different slopes in the temperature range 300K - 700K. According to Mott and Davis [7], contribution to the conduction due to carriers hopping at the Fermi level becomes dominent at rather low temperatures not covered in our experiment. Based on these arguments the linear region I could be attributed to the conduction through the extended bands and the region II to hopping at the band edges.

References:

01. J.L. Shay and J.H. Wernick, Ternary Chalcopyrite Semiconductors; Growth, Electronic properties and applications. Oxford, Pergamon Press, (1975).

02. W. Honig, H. Neumann, H. Sobotta, B. Schumann and G. Kuhn; Thin solid films 48, 67, (1978).

03. L.Y. Sun, L.L. Kazmerski, A.H. Clark, P.J. Ireland and D.W. Mortan; J. Vac. Sci. Tech. 15, 265, (1978).

04. H. Neumann, B. Perlt, N.A.K. Abdul-Hussein, R.D. Tomlinson and A.E. Hill; Crystal Res. Tech. 17, 5 (1982).

05. Solar Cells Journal vol. 16 (1986); Special Issue on $CuInSe_2$.

06. J.L. Shay, S. Wagner, H.M. Kasper; Appl. Phys. Lett. 27, 89, (1975).

07. N.F. Mott and E.A. Davis; Electronic process in non crystalline materials; Oxford, Clarendon Press, (1979).

08. L.J. Van der Pauw; Philips Res. Reports 13, 1-9, (1985).

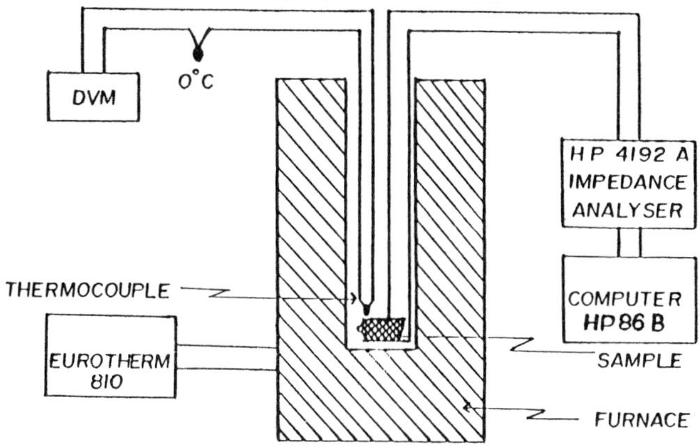

Fig. 1. A schematic diagram of the arrangement used for the measurement of ac conductivity of $CuInSe_2$.

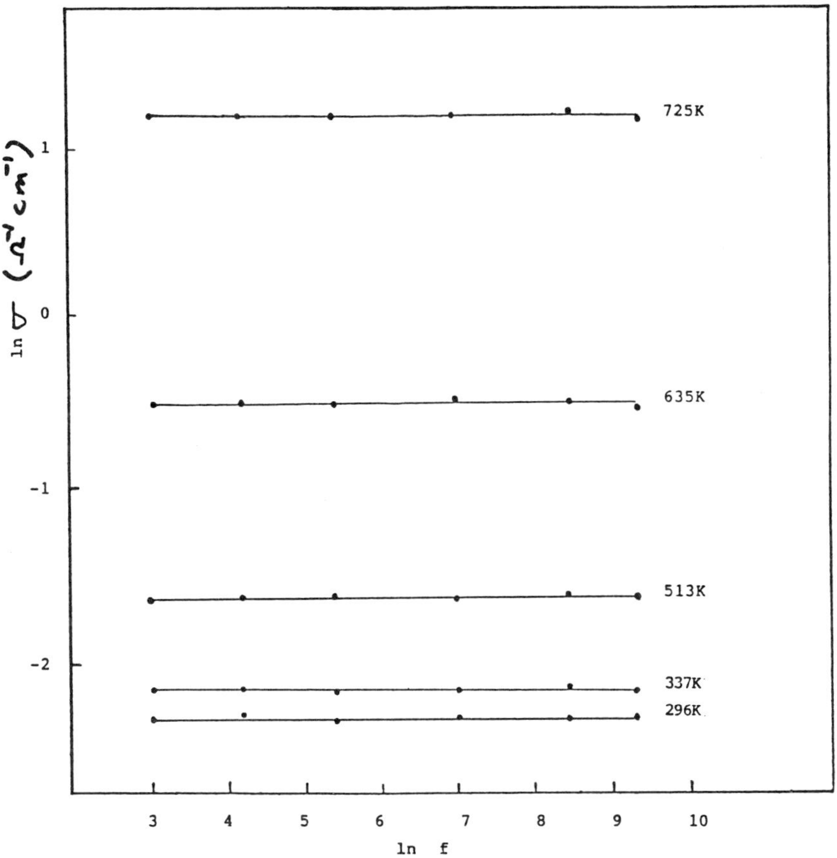

Fig. 2. Frequency dependence of ac conductivity for $CuInSe_2$.

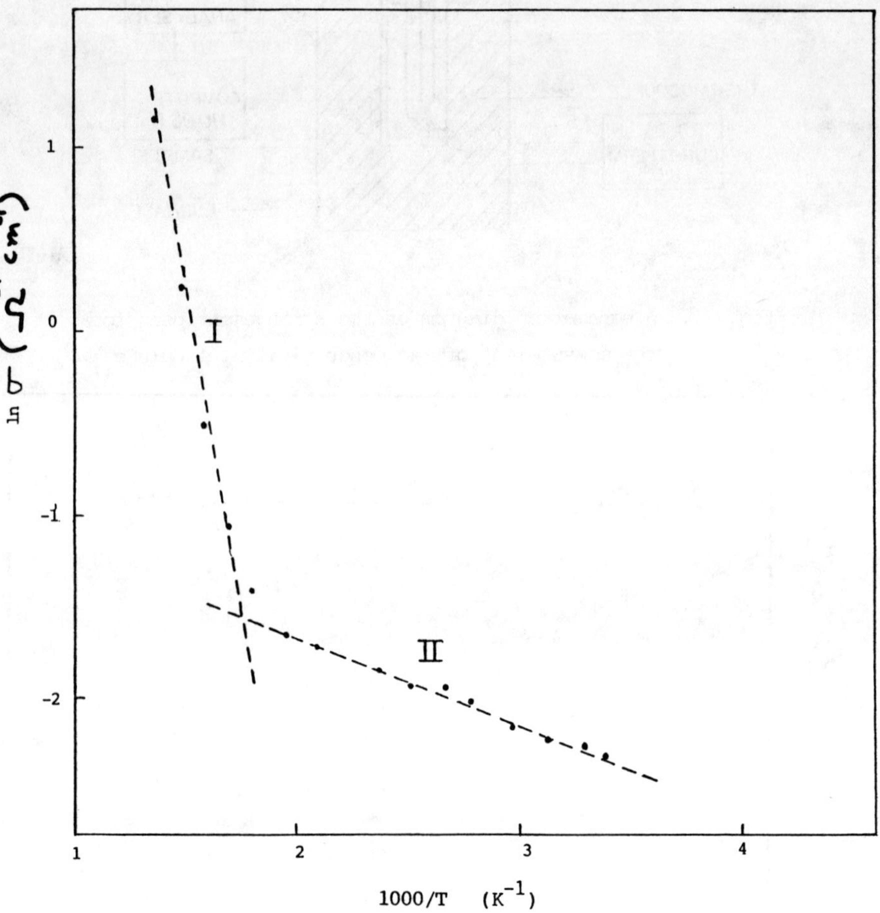

Fig. 3. Temperature dependence of dc conductivity of $CuInSe_2$.

BAND-SPLITTING IN CUBIC LATTICES WITHIN THE METAL-INSULATOR TRANSITION MODEL

M. Ahmed
Department of Physics
University of Dhaka
Dhaka 2, Bangladesh

Abstract: Transition-metal and rare-earth oxides, sulfides, and borides exhibit an unusually wide variety of electrical and magnetic properties. Some of these substances show a large variation in resistivities with temperature and present a first-order transition in resistivity as a function of temperature. Falikov-Kimball gave a simple theoretical model for such metal-insulator transitions. The model was first solved in the mean-field approximation by Ramirez et al. Later works attempted to solve the model using coherent-potential approximation, but results are in conflict with each other.

In this context we solve the model within the electron correlation approximation of Zubarev and obtain the single particle Green's function. For nonvanishing interaction of localized and band states G, we obtain splitting of the band which diminishes for a large value of the interaction. We calculate the number of conduction electrons as a function of temperature and the thermodynamic potential for a simple cubic lattice. In general, these are weakly dependent on G and we find a split band situation for small bandwidth W and for large W there is only one band. In both cases we only find the insulating phase and no metallic phase or phase transitions of any kind, in agreement with the calculations of Ghosh.

Transition-metal as well as rare-earth oxides, sulfides, and borides exhibit an unusually wide variety of electrical and magnetic properties (Bhide and Dani 1961; Ferretti et al. 1965; Adler 1968; Balseiro et al. 1980; and Balseiro and Falicov 1985). Some of them are excellent conductors, some are insulators, while some substances show a large variation in resistivities with temperature (Abrahams and Calhoun 1953; Gorkov and Lebed 1984). The last type of materials present a first-order transition in resistivity as a function of temperature. Falicov and Kimball (1969) gave a simple theoretical model for such metal-insulator transitions. The model is based on the existence of both localized (ionic) and band (Bloch) states. The electron-hole interaction was found to be responsible for the anomalous temperature dependence of the conduction electrons.

In the language of second quantization the model hamiltonian was written as

$$\mathcal{H} = \sum_{\nu k\sigma} \epsilon_\nu(\vec{k})\, a^+_{\nu\vec{k}\sigma}\, a_{\nu\vec{k}\sigma} + \sum_{i\sigma} E b^+_{i\sigma} b_{i\sigma} - G \sum_{i\sigma\sigma'\nu\nu'} b^+_{i\sigma} C_{\nu i\sigma'} C_{\nu' i\sigma'} b_{i\sigma} \tag{1}$$

where $a^+_{\nu\vec{k}\sigma}$ creates an electron in the state $\vec{k}$ with band ν with spin σ and $b^+_{i\sigma}$ creates a hole with spin σ at site i.

Here $G > 0$ corresponds to an attractive electron-hole interaction independent of spin and band index. The operator $C^+_{\nu\vec{i}\sigma}$ creates a Wannier state at site i corresponding to the Bloch band $(\nu\sigma)$ and is related to $a^+_{\nu k\sigma}$ by the transformation

$$C^+_{\nu j\sigma} = N^{-1/2} \sum_{\vec{k}} e^{i\vec{k}\cdot\vec{R}_j}\, a^+_{\nu k\sigma} \tag{2}$$

The energies $\varepsilon_\nu \vec{k})$ and E are positive and definite such that

$$\Delta = \min[E + \epsilon_\nu(\vec{k})] > 0 \tag{3}$$

The model was first solved in the mean field approximation (Ramirez et al. 1970) for bandwidth $W = 0$ and $W \neq 0$. Depending on the relative values G/Δ and W/Δ in this approach, it was possible to obtain different behaviors as a function of temperature: a pure insulator, pure metal, or metal-insulator transitions. later works (Plischke 1972; Goncalves da Silva and Falicov 1972; Heritier et al. 1984; Coleman et al. 1985) have attempted to solve the model using coherent-potential approximation (Velicky et al. 1968; Hubbard 1964b; Friedel 1985). Unfortunately, there is no agreement between these papers.

Plischke solved the model using coherent-potential approximation. Using an unperturbed simple cubic density of states, he obtained insulating and metallic states but did not find any transition. Goncalves da Silva and Falicov took a square density of states and by introducing extra approximations relating to the band edges and shape, obtained qualitatively the same kind of behavior as that of Ramirez et al. Finally, Ghosh (1976), employing the same density of states as Plischke found only the insulating phase with no metallic phase or phase transition or any kind.

In this context, we believe it can be useful to solve the model within the electron correlation approximation proposed by Zubarev (1960) and used by Hubbard (1963, 1964a, 1964b) and many others. Our calculations show band splitting for any nonvanishing G and for a large $2G/W$ value this splitting diminishes.

Following Zubarev we define the retarded Green's function

$$\langle\langle C_{\nu k\sigma}(t);\, C^{+}_{\nu k\sigma}(t')\rangle\rangle = -i\theta(t - t')\,\langle[C_{\nu k\sigma}(t);\, C^{+}_{\nu k\sigma}(t')]\rangle \tag{4}$$

where $[c, c^{+}] = cc^{+} - c^{+}c$ and $\theta(x)$ is a step function, i.e., $\theta(x) = 1$ if $x > 0$ or zero otherwise. The Green's function equation of motion can be written as

$$i\frac{d}{dt}\langle\langle C_{\nu k\sigma}(t);\, C^{+}_{\nu \vec{k}\sigma}(t')\rangle\rangle = \delta(t - t')\langle[C_{\nu \vec{k}\sigma}(t);\, C^{+}_{\nu \vec{k}\sigma}(t')]\rangle$$
$$\langle\langle[C_{\nu \vec{k}\sigma}(t)\mathcal{H};\, C^{+}_{\nu \vec{k}\sigma}(t')\rangle\rangle \tag{5}$$

Since the Green's function is a function of $t - t'$ only, we can define, for real quantities, the Fourier transforms

$$\langle\langle C_{\nu \vec{k}\sigma}(t);\, C^{+}_{\nu \vec{k}\sigma}(t')\rangle\rangle_{\omega} = \frac{1}{2\pi}\int \langle\langle C_{\nu \vec{k}\sigma}(t);\, C^{+}_{\nu \vec{k}\sigma}(0)\rangle\rangle e^{i\omega t} dt \tag{6}$$

and hence the equation of motion becomes

$$\epsilon_k\,\langle\langle C_{\nu \vec{k}\sigma};\, C^{+}_{\nu \vec{k}\sigma}\rangle\rangle = \frac{1}{2\pi}\langle[C_{\nu \vec{k}\sigma};\, C^{+}_{\nu \vec{k}\sigma}]\rangle + \langle\langle[C_{\nu \vec{k}\sigma}, \mathcal{H};\, C^{+}_{\nu \vec{k}\sigma}\rangle\rangle_{\omega} \tag{7}$$

Using the decoupling procedure (Zubarev 1960) we can solve for the Green's We obtain

$$\langle\langle C_{\nu \vec{k}\sigma};\, C^{+}_{\nu \vec{k}\sigma}\rangle\rangle = \frac{1}{2\pi}\left(\frac{A_1}{\omega - \omega_1} - \frac{A_2}{\omega - \omega_2}\right) \tag{8}$$

$$A_i = \frac{[\omega_i - \epsilon_0 + G(1 - n_h)]}{(\omega_1 + \omega_2)} \tag{9}$$

$$\omega_i = \frac{1}{2}(\epsilon_0 - G + \varepsilon_k) \pm \sqrt{\left(\frac{\varepsilon_0 - G - \varepsilon_k}{2} - Gn_h(\varepsilon_k - \varepsilon_0)\right)} \tag{10}$$

In these expressions, $n_h = \Sigma\langle b^{+}_{i\sigma} b_{i\sigma}\rangle$ is the mean value of the hole population in every localized site, ε_k is the band dispersion relation, and ε_0 is the single particle energy of band electrons.

We now calculate the number of conduction electrons as a function of temperature, $n_e(T)$. The grand partition function, Z and the thermodynamic potential Ω are related by

$$\Omega = - KT\ell nZ.$$

By writing the Hamiltonian as $H = H_0 + \lambda H_1$ where λH_1 is the hopping term, ($\lambda_M = \varepsilon_\nu(k)$ = const) it is possible (Schultz 1969) to express Ω as

$$\Omega = \Omega_0 + \int^{\lambda_m} \frac{d\lambda}{\lambda}\langle \lambda H_1\rangle_\lambda$$

where Ω_0 is the solution for $W = 0$ and the exact grand partition function is found to be ($\mu_0 = \mu_e = -\mu_h$)

$$\begin{aligned} Z_0 = \Big\{ 1 &+ 2\exp\left[-\beta(\varepsilon_0 - \mu_0)\right] + \exp\left[-2\beta(\varepsilon_0 - \mu_0)\right] \\ &+ 2\exp\left[-\beta(E + \mu_0)\right] + 4\exp\left[-\beta(\varepsilon_0 + E + G)\right] \\ &+ 2\exp\left[-\beta(2\varepsilon_0 + E + 2G - \mu)\right]\Big\}^N \end{aligned} \tag{11}$$

N is the number of sites. The chemical potential μ_0 can be determined by setting $(h_h)_0 = (\dot{n}_e)_0$.

From general relations of statistical mechanics, we find $\partial\Omega/\partial\mu = \langle n_h - n_e\rangle$ and $\partial^2\Omega/\partial\mu^2 < 0$. In our case we find $\langle n_h - n_e\rangle = 0$, thus in practice we have different values for n_h; from here we find different μ and the corresponding Ω. We can verify this procedure in the limit $W = 0$, by comparing results found for n_e and μ from the extremum of Ω as obtained from the Green's function with the exact result deduced from Z_0.

Finally, we calculate the thermodynamic potential for a simple cubic lattice with energy band given by

$$\varepsilon_k = \varepsilon_{k0} + \frac{W}{6}(\cos\varepsilon_x a + \cos\varepsilon_y a + \cos\varepsilon_z a)$$

where a is the lattice constant. We obtain

$$\Omega = N\left[\Omega_0 + \int_0^{W/12} \frac{d\lambda}{\lambda} \sum_{\delta\neq 0(n,n)} \lambda \langle G^+_{k+\delta,\sigma}, C_{\vec{k}\sigma}\rangle_\lambda\right] \tag{12}$$

Using the Fourier transform

$$\langle G^+_{k+\delta,\sigma}, C_{\vec{k}\sigma}\rangle_\lambda = \frac{1}{N}\sum_q e^{i\vec{q}\cdot\vec{a}} \langle G^+_{\vec{k},\sigma}, C_{\vec{k}\sigma}\rangle$$

the thermodynamic potential Ω becomes

$$\Omega = N\left\{\Omega_{ok} + \frac{4}{N}\sum(\cos\varepsilon_x a + \cos\varepsilon_y a + \cos\varepsilon_z a) \times \int^{W/12} d\lambda \left[f(\omega_1)A_1 + f(\omega_2)A_2)\right]_\lambda\right\} \tag{13}$$

with

$$f(\omega) = \left\{\exp\left[\beta(\omega - \mu) + 1\right]\right\}^{-1} \tag{14}$$

The sums over the reciprocal vectors k are taken over the first Brillouin zone.

In Figure 1 we plot the number of carriers as a function of reduced inverse temperature. In general, it can be said that we find very little G-dependence. For $W = 7\Delta$ we find a split band situation as predicted by the coherent-potential approximation method while for $W = 70\Delta$ there is only one band. In both cases, we only find the insulating phase. Therefore our results agree qualitatively with that of Ghosh (1976).

Acknowledgement

The author is grateful to Professor Sir Sam Edwards for his warm hospitality at the Cavendish Laboratory and for many stimulating discussions.

References

1. Abrahams, R. F., Calhoun, B. A., *Acta Cryst.*, vol. 6, p. 105, 1953.
2. Adler, D., *Rev. Mod. Phys.*, vol. 40, p. 714, 1968.
3. Balsiero, C. A., Schlottmann, P, Yndurain, F., *Phys. Rev. B*, vol. 21, p. 5267, 1980.
4. Balseiro, C. A., Falicov, L. M., *Phys. Rev. Lett.*, vol. 55, p. 2336, 1985.
5. Bhide, V. G., Dani, R. H., *Physica*, vol. 27, p. 821, 1961.
6. Coleman, R. V., Eiserman, G., Everson, M. P., Johnson, A., Falicov, L. M., *Phys. Rev. Lett.*, vol. 55, p. 863, 1985.
7. Ferretti, A., Rogers, D. B., Goodenough, J. B., *J. Phys. Chem. Solids*, vol. 26, p. 2007, 1965.
8. Freidel, J., *Philos. Trans. Roy. Soc. London*, vol. A314, p. 189, 1985.
9. Goncalves da Silva, C. E. T., Falicov, L. M., *J. Phys. C: Solid St. Phys.*, vol. 5, p. 906, 1972.
10. Gor'kov, L. P., Lebed, A. G., *J. Phys. Lett.*, vol. 45, p. 433, 1984.
11. Heritier, M., Montambaux, G., Lederer, P., *J. Phys. Lett.*, vol. 45, p. 943, 1984.
12. Hubbard, J., *Proc. Roy. Soc. A*, vol. 276, p. 238, 1963.
13. Hubbard, J., *Proc. Roy. Soc. A*, vol. 277, p. 237, 1964a.
14. Hubbard, J., *Proc. Roy. Soc. A*, vol. 281, p. 401, 1964b.
15. Plischke, M., *Phys. Rev. Lett.*, vol. 28, p. 361, 1972.
16. Ramirez, R., Falicov, L. M., Kimball, J. C., *Phys. Rev. B*, vol. 2, p. 3383, 1970.
17. Schultz, T. D., *Quantum Field Theory and the Many Body Problem*, p. 134, New York: Gordon and Breach, 1969.
18. Velicky, B., Kirkpatrick, S., Ehrenreich, H., *Phys. Rev.*, vol. 175, p. 747, 1968.
19. Zubarev, D. N., *Usp. Fiz. Nauk*, vol. 71, 1960; translation from *Sov. Phys. Uspekhi*, vol. 3, p. 320.

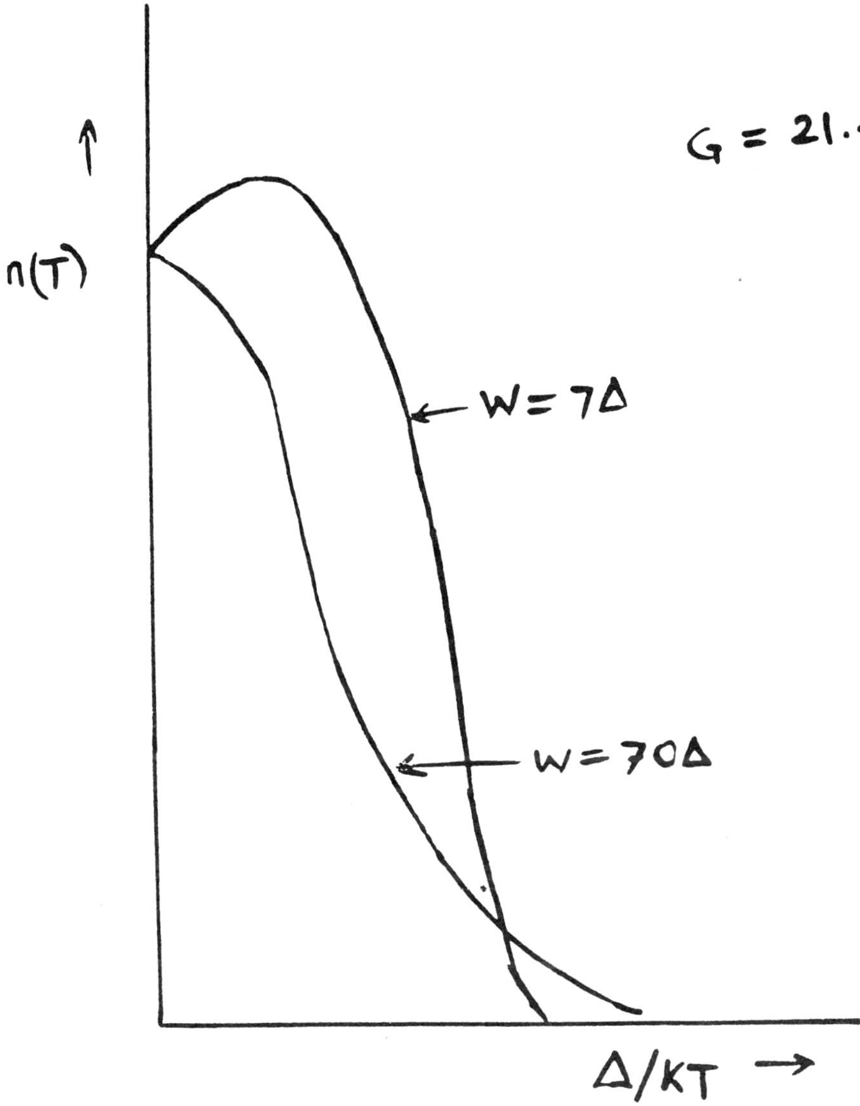

Fig. 1. Number of carriers, n as a function of reduced inverse-temperature Δ/kT; for $G = 21.4\ \Delta$.

SUBSOLIDUS PHASE EQUILIBRIA IN THE TERNARY SYSTEM CaNaPO4-Ca2 SiO4-Ca3(PO4)2

R.P. Gunawardana
Department of Chemistry, University of Peradeniya
Peradeniya, Sri Lanka

Introduction:

Rhenanite, $CaNaPO_4$ is considered as a commercially important phosphate fertilizer[1]. Rock phosphate can be converted to rhenanite by the solid state reaction[2] of rock phosphate with soda ash and quartz. Since the relevant bulk compositions lie in or close to the ternary system $CaNaPO_4$-Ca_2SiO_4-$Ca_3(PO_4)_2$, the subsolidus phase equilibria of this system has been determined in order to obtain the optimum composition and other conditions suitable for the conversion of rock phosphate to rhenanite.

Experimental:

Ca_2SiO_4 and $CaNaPO_4$ were prepared from Na_2CO_3, $CaCO_3$, $CaHPO_4.2H_2O$ and pure crushed quartz. Phase equilibria were determined by preparing mechanical mixtures of the three phases Ca_2SiO_4, $CaNaPO_4$ and $\beta Ca_3(PO_4)_2$ in the desired proportions and sintering them for 1-4 days at constant temperature. All preparations were made in Pt foil envelopes or crucibles and heated in muffle furnaces. The data were obtained on three isothermal sections at 1300°C, 1100°C and 950°C. The reaction products were quenched and the phases present were identified by powder X-ray diffraction using CuK_α radiation.

Results and Discussion:

The data on the binary edges were obtained from the literature[3,4,5]. Isothermal sections at 1300°, 1100° and 950°C have been constructed from the phases present in annealed compositions and from the available

FIGURE 1.

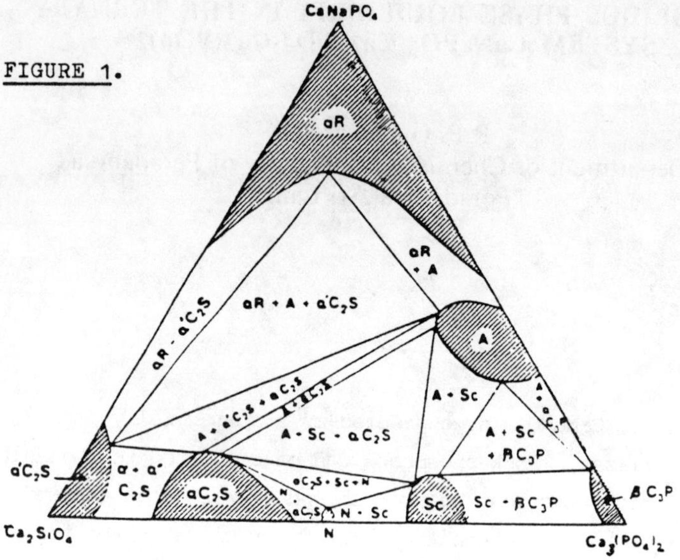

1100° C isothermal section of the system $Ca_3(PO_4)_2$–Ca_2SiO_4–$CaNaPO_4$ (in wt.%).

data on the binary systems. Isothermal sections at 1100°C and 1300°C are shown in Figures 1 and 2 respectively. 950°C isotherm is very similar to the one at 1100°C and as such it is not given separately.

Isothermal sections are dominated by several series of extensive solid solutions. The field of homogeneous α-rhenanite solid solution reaches its maximum extent along the two binary edges but within the ternary system its extent is less. This concave limit suggests that the maximum solubility in $CaNaPO_4$ or either Ca_2SiO_4 or $Ca_3(PO_4)_2$ is depressed by the presence of the other component. The solid solutions thus have a strong negative deviation from ideality.

The field of phase A, $CaNaPO_4.Ca_3(PO_4)_2$, has a large extent towards SiO_2-rich compositions, especially at 1300°C. Nagelschmidtite, $Ca_7(PO_4)_2(SiO_4)_2$, appears only on the 1100°C isotherm and its composition is essentially fixed. Silicocarnotite, $Ca_5(SiO_4)(PO_4)$, α'Ca_2SiO_4 and αCa_2SiO_4 have progressively larger fields of single phase homogenity.

FIGURE 2.

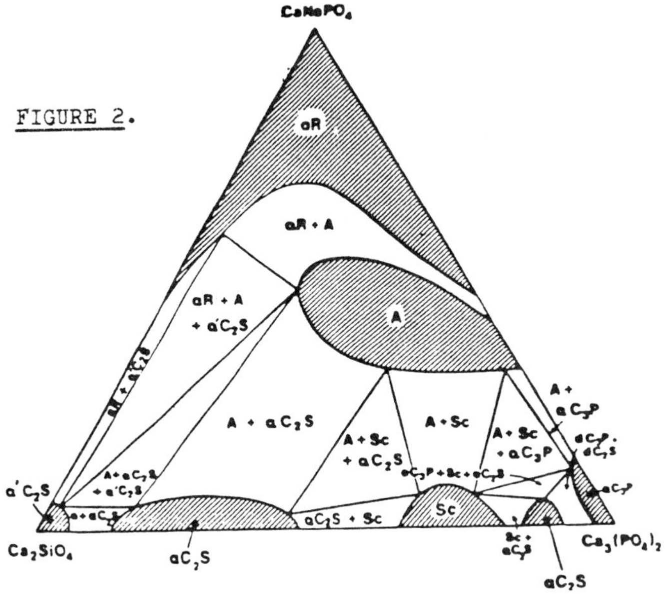

1300° C isothermal section of the system $Ca_3(PO_4)_2$–Ca_2SiO_4–$CaNaPO_4$ (in wt.%).

Two ranges of α Ca_2SiO_4 solid solution are stable at 1300°C, one lies at low $Ca_3(PO_4)_2$ contents while the other is much richer in the $Ca_3(PO_4)$ component. There are several two-phase and three-phase regions marked in the isothermal sections. No new phases have been observed in this ternary section.

Contrasting the two isothermal sections (at 1100° and 1300°), a marked increase in solid solution occurs with rising temperatures and the two phase regions of phase A - rhenanite and phase A - α Ca_2SiO_4 both expand rapidly over this temperature interval. On both isothermal sections, phase A solid solutions are in equilibrium with all the other phases.

These phase equilibrium data coupled with solubility studies of the phases present in annealed samples revealed optimum compositions and other conditions (such as temperature etc.) most suited for the formation of rhenanite solid solution, avoiding the phases which are considered undesirable in a P-fertilizer.

References:

1. F. Ullmann, *Éncyclopadie der Technischen Chemie*, Vol.6 (Urban and Schwarzenberg, Munchen, Berlin, Wien, 1955).

2. R.P. Gunawardane, *Proc. Sri Lanka Assoc. Advmt. Sci.*, 35, 56, (1979).

3. J. Ando and S. Matsuno, *Bull. Chem. Soc. Japan*, 41, 342, (1968).

4. R.W. Nurse, J.H.Welch and W.Gutt, *J. Chem. Soc*. 1078, (1959).

5. V.M. Ustyantsey and M.G.Tretnikova, Izv. *Akad. Nank SSSR, Neorg. Mater*, 13, 865 (1977).

DIELECTRIC PROPERTIES OF BISMUTH-BORATE GLASSY SOLID ELECTROLYTES WITH A SPECIAL REFERENCE TO THEIR ALKALI CONTENT

K. Singh, R. Kahandewal
Department of Physics
Nagpur University

Abstract: The dielectric properties of bismuth-borate glass systems of series I ($60B_2O_3$ - 40 Bi_2O_3) and series II ($80B_2O_3$ - $20Bi_2O_3$) with Li_2O addition at the cost of Bi_2O_3, have been investigated as the function of temperature and frequency in the temperature range from room temperature to 400°C and in the frequency range from $5{\cdot}10^4$ Hz to $3{\cdot}10^7$ Hz, respectively. As Li content increases in these series, dielectric constant increases rapidly with the rise in temperature. The frequency dependence undergoes a relaxation and carries diffusion process due to increased density of Li Ion charge species.

1. Introduction

Conductivity measurements alone are not sufficient to study the mechanism of ion displacement. Other properties, such as low-frequency dielectric relaxation or self-diffusion of charge carriers, occur in glass through displacement of these ions and can contribute to an understanding of ionic conduction mechanism in vitreous phases. Modulus spectroscopy was first introduced by glass scientists as a mathematical tool for analyzing conductivity relaxation in glasses and in concentrated aqueous solutions [1].

In the presence of relaxation effects, the two frequency dependent dielectric constants ε' and ε'' are combined into a single complex dielectric constant $\varepsilon^* = \varepsilon' - j\varepsilon''$ and the phase angle δ between ε' and ε'' is given by $\tan\delta = \varepsilon''/\varepsilon'$. Oxide glasses, containing alkali metal oxide are ionic conductors. The electric current is carried entirely by alkali metal ions as has been shown by several authors [2-4].

Namikawa [5] has reported these types of diffusion processes defined by the correlation between conduction and dielectric relaxation; the localized diffusion process causing the Maxwell-Wagner loss, the nonlocalized diffusion process I causing migration loss I and conduction I, and nonlocalized diffusion process II causing migration loss II and conduction II. One or more of these diffusion processes can be observed in all oxide glasses.

The present paper describes the attempt to extend the above concept to series I and III with $60B_2O_3$-$40Bi_2O_3$ and $80B_2O_3$-$20Bi_2O_3$, respectively. In these series Li_2O is added at the cost of Bi_2O_3. The dielectric constant and loss tangent have been measured as the function of temperature and frequency.

It has already been reported by us that as LiO_2 content increases electrical conductivity increases with a rise in temperature [6].

The present investigation is undertaken to set up a correlation of electrical conduction with dielectric relaxation for understanding the electrical properties of bismuth-borate glasses with a special reference to alkali ion concentration.

2. Experimental

The glasses in series I ($60B_2O_3$-$40Bi_2O_3$) and series II ($80B_2O_3$-$20Bi_3O_3$) with the varying concentration of LiO_2 were prepared from reagent grade chemicals. The compositions are listed in Table I (a, b). The ingredients were thoroughly mixed under acetone and dried. The dried mixture was heated in a porcelain crucible at a temperature 293 K above the melting point and maintained for one hour to homogenize the melt. The melt was then quenched in an aluminum mould at room temperature to obtain glasses of the desired shape. The quenched samples were annealed at

Table 1a

Composition of glasses in series I

Glass No.	B_2O_3 (mole %)	Bi_2O_3 (mole %)	Li_2O (mole %)
I - 0	60	40	0
II - 1	60	30	10
III - 2	60	20	20
IV - 3	60	10	30
V - 4	60	0	40

Table 1b

Composition of glasses in series II

Glass No.	B_2O_3 (mole %)	Bi_2O_3 (mole %)	Li_2O (mole %)
I - 0	80	20	0
II - 1	80	15	5
III - 2	80	10	10
IV - 3	80	5	15
V - 4	80	0	20

423 K for two hours. The glass plates were polished using silicon carbide grit of different mesh sizes to obtain samples with dimensions of 8 mm × 5 mm × 2.5 mm. To achieve good ohmic contact, two flat surfaces of the glass were coated with a thin layer of Ag paint. Samples were baked in an oven at 573 K for four hours to ensure perfect adhesion.

The dielectric constant of these samples was measured as the function of temperature and frequency in the temperature range RT to 623 K and frequency $5 \cdot 10^4$ Hz to $3 \cdot 10^7$ Hz using USSR Q-meter (type E 9-4).

3. Results and Discussion

The functional dependence of the dielectric constant on temperature for series I and II (Figures 1a and b) shows that the dielectric constant increases exponentially with temperature. The increase in dielectric constant with increasing Li content can also be clearly seen from these curves.

At higher temperatures there is an increasing contribution resulting from ion mobility which in turn increases the dielectric constant remarkably.

With increasing alkali content the decrease in the dielectric constant at higher temperatures may be attributed to the small size of the Li ion jump frequency for ion motion resulting in an increased concentration of charge carriers.

Figures 2a, b, and 3a, b show the variation of dielectric constants ε and $\tan\delta$ of series I and II with frequency. These figures demonstrate that the alkali ion content affects dielectric constant ε and $\tan\delta$ considerably. Both ε and $\tan\delta$ undergo resonance and antiresonance phenomena. Two resonance peaks appear in ε and $\tan\delta$ denoted by I and II in the frequency range from 10^{15} to 10^{17} Hz.

From Figures 2a and b it is clear that as alkali ion content increases $_{\text{peak}}$I and $_{\text{peak}}$II continue to be modified.

The occurrence of the two relaxation peaks at the frequency dependence of the dielectric constant and the frequency dependence of $\tan\delta$ can be explained by assuming that two types of diffusion process: localized and nonlocalized carrier, occur in the mixed alkali glasses.

The occurrence of peak I in the lower frequency region is attributed to the nonlocalized carrier diffusion process. This process arises because of the movement of charge carriers. This movement is restricted to a zigzag path turning at random directors. Under an applied field, these charge carriers make a transition from one stable side to another. The migration gives rise to conduction current.

The migration of charge carriers from one site to another gives rise to resonance absorption at the lower frequency region.

The presence of peak II is attributed to a localized diffusion process. This process arises because of the presence of materials with different conductivities. These inhomogeneous are randomly distributed in the sample.

This processes arises because of the presence of significant densities of localized electronic and ionic carriers. When an a.c. field is applied across the sample, these carriers are capable of random hopping between specially fixed centers. This process gives rise to resonance absorption of energy in the high frequency region. This is usually called the Maxwell-Wagner loss.

The decrease in the magnitude of the Maxwell-Wagner loss, when the Li_2O content is increased in series I and the near disappearance of this loss when the Li_2O content is 40 mol-% shows that the localized diffusion process changes to the nonlocalized process in this case.

4. Conclusion

1. There are two types of diffusion processes in the present case defined by the correlation between conduction and dielectric relaxation; localized and nonlocalized carrier.

2. It can be assumed that the microstructure of glass might affect the movement of mobile carriers determining the diffusion process in oxide glasses. This is reflected in series I and II.

3. The transition of the diffusion process from one type to another with the change of Bi_2O_3 and Li_2O compositions explains the sensitivity of two types of diffusion process in these glasses.

References

1. Macedo, P. B., Mohnihan, C. T., Bose, R., The Role of Ionic Diffusion in Polarization in Vitreous Ionic Conductors, *Phys. Chem. Glasses*, vol. 14, p. 171, 1971.
2. Kraus, C. A., Darby, E. J., *J. Amer. Chem. Soc.*, vol. 44, p. 2783, 1922.
3. Evstropev, K. K., Pavlova, G. A., *Trudy Leingr. Tekh. Inst. Im Lensovbeta*, vol. 46, p. 49, 1958.
4. Hughes, K., Israd, J. O., *Phys. Chem. Glasses*, vol. 9, p. 37, 1968.
5. Namikawa, H., *J. Non-cryst. Solids*, vol. 18, p. 173, 1975.
6. Singh, K., Kahandewal, R., Deshpande, V. K. *Mats. for Sol. St. Batteries*, Ed. by B. V. R. Chowdari and S. Radhakrishna, Singapore: World Scientific Publ. Co., p. 405, 1986.

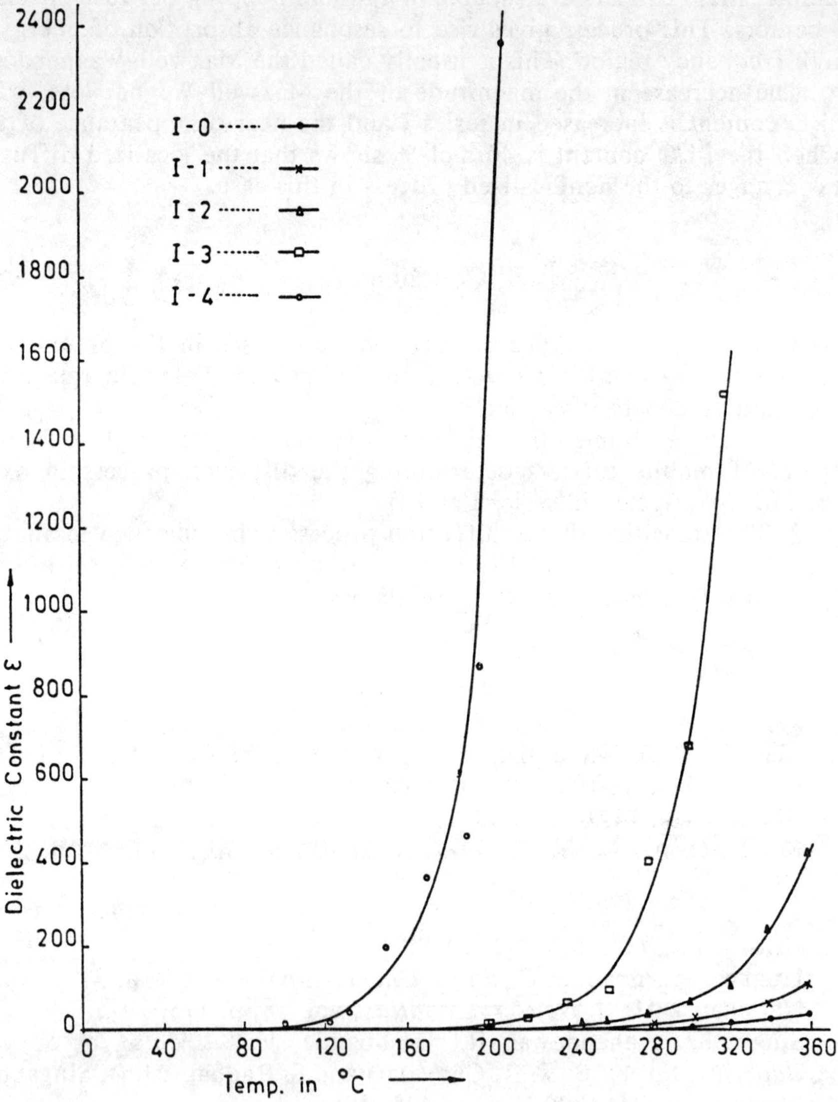

Fig. 1a. Temperature dependence of dielectric constant ε for series I.

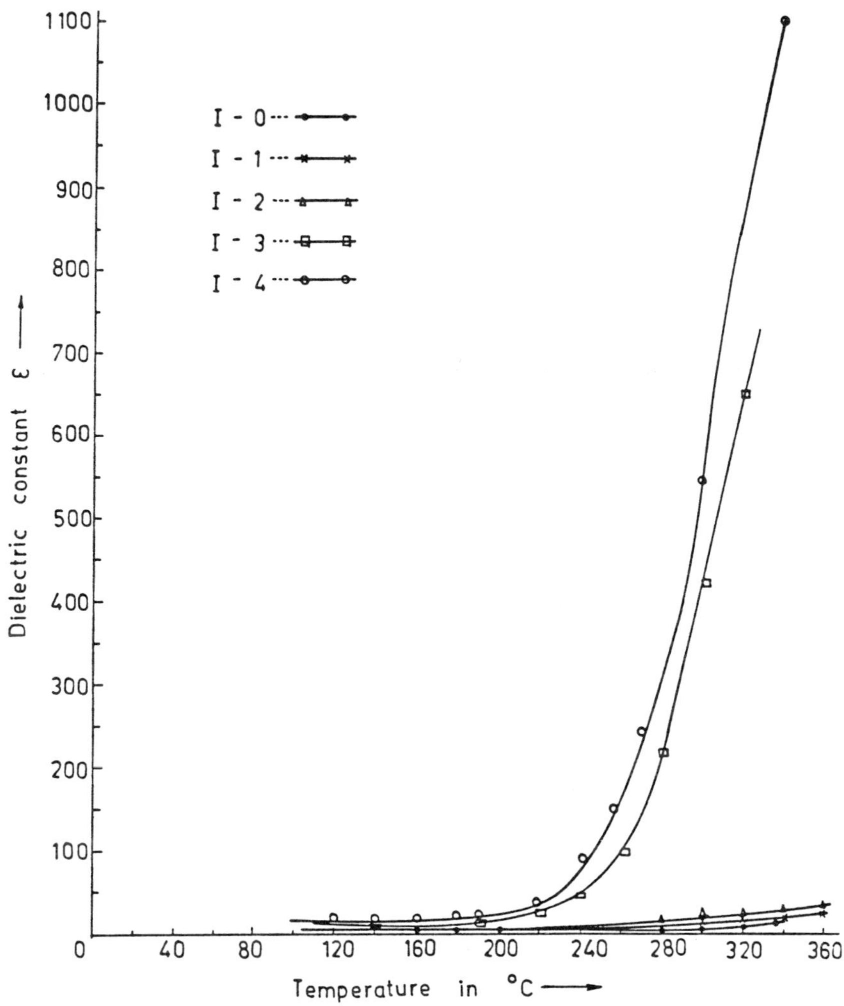

Fig. 1b. Temperature dependence of dielectric constant for series II.

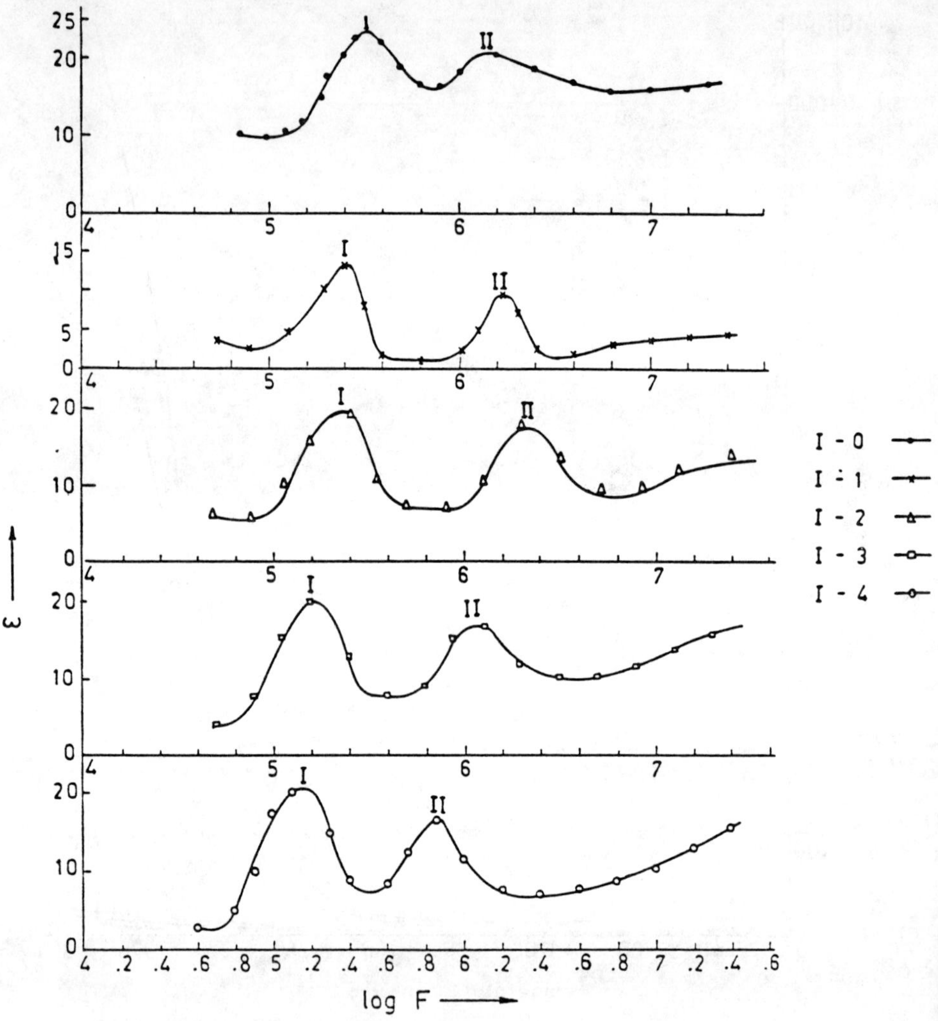

Fig. 2a. Frequency dependence of dielectric constant at room temperature for series I.

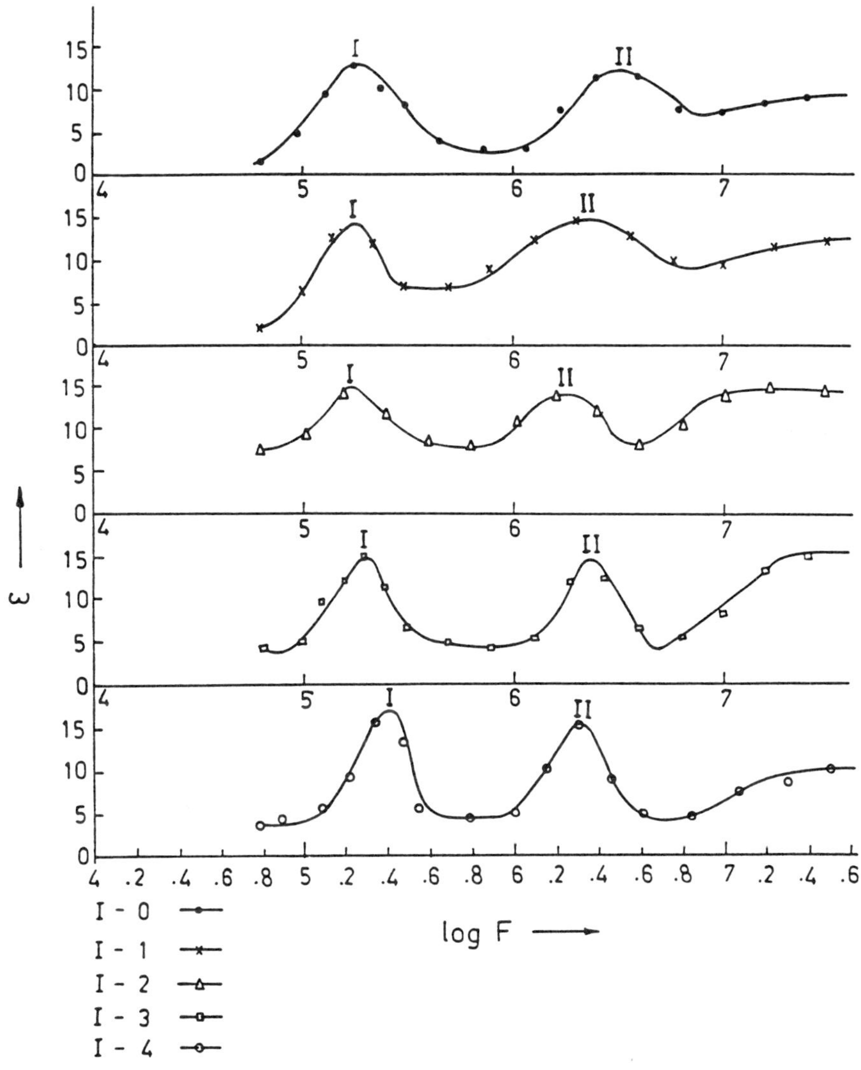

Fig. 2b. Frequency dependence of dielectric constant at room temperature series II.

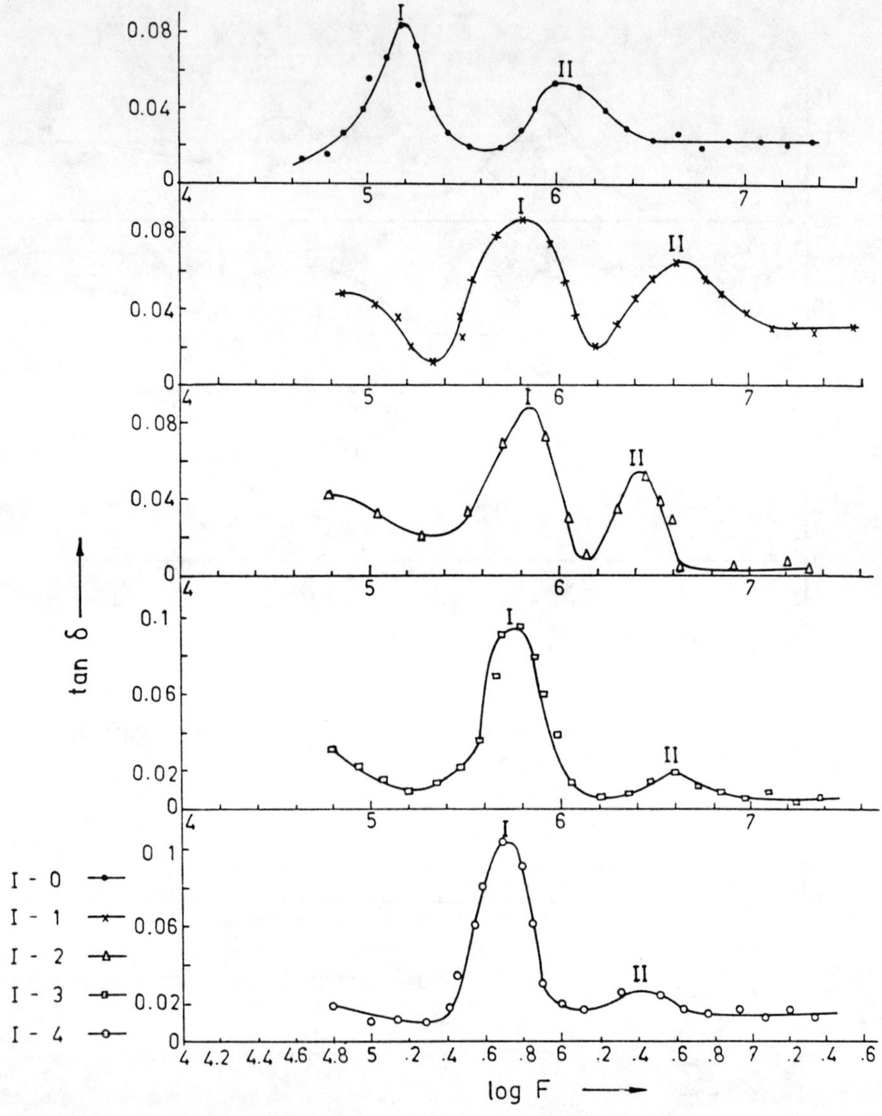

Fig. 3a. Frequency dependence of loss tangent at room temperature for series I.

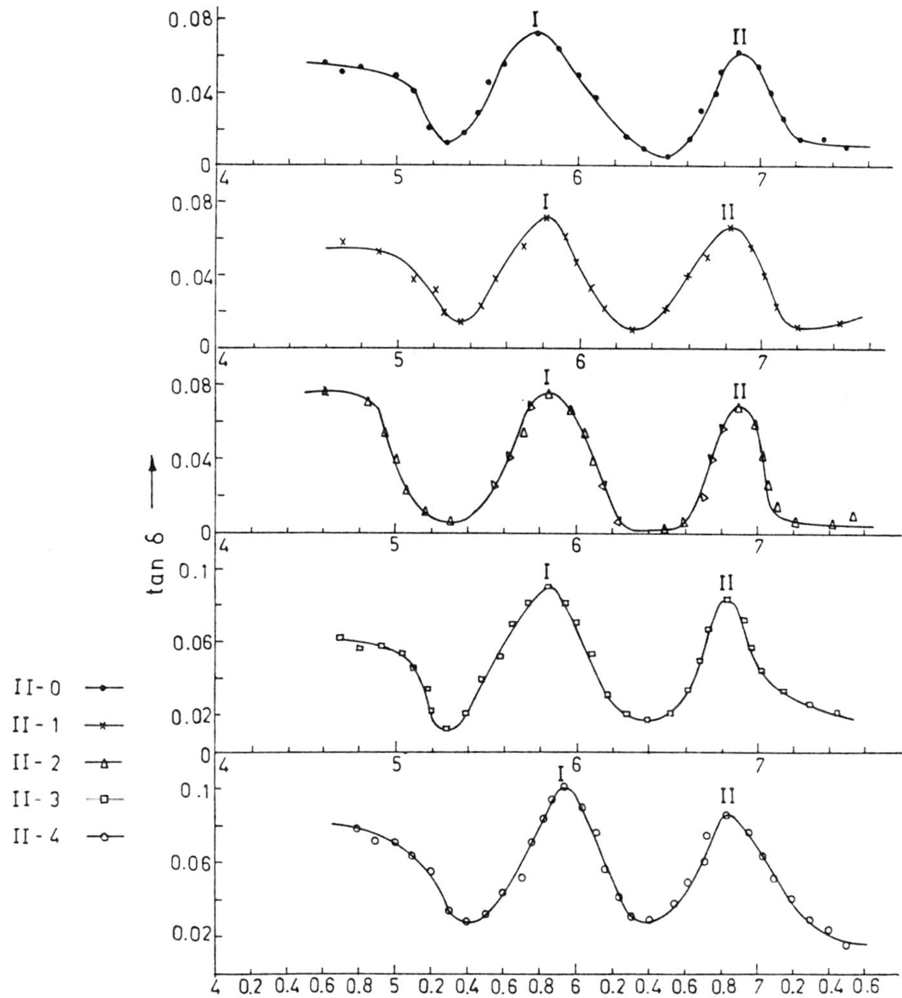

Fig. 3b. Frequency dependence of loss tangent at room temperature for series II.

ELECTRICAL CONDUCTION OF DITHIOCARBAMATE COMPLEXES OF MOLYBDENUM (III, IV), IRON (III) AND CHROMIUM (III)

H. M. N. Bandara, O. A. Ileperuma, P. A. G. D. Dharmasena
Department of Chemistry, University of Peradeniya
Peradeniya, Sri Lanka

M. A. K. L. Dissanayake
Department of Physics, University of Peradeniya
Peradeniya, Sri Lanka

Abstract: **The d.c. electrical conductivity of dimethyl-dithiocarbamate complexes of molybdenum (III), molybdenum (IV), iron (III), and chromium (III) have been measured using pressed pellets. Molybdenum (III) and molybdenum (IV) dithiocarbamates are observed to have conductivities of $2.5 \cdot 10^{-3}$ $ohm^{-1}cm^{-1}$, respectively, whereas iron (III) and chromium (III) complexes are poor conductors with conductivities less than 10^{-6} ohm^{-1} cm^{-1} at room temperature. An attempt has been made to explain these observations in terms of structural differences of the compounds.**

Introduction:

There is a growing interest in the physical and chemical properties of dithiocarbamate complexes (1,2) of molybdenum and iron due to the possibility of these being used as model compounds for biological enzyme systems. Nitrogenase, aldehyde oxidase, nitrate reductase and sulphite oxides are some of the enzymes containing molybdenum and/or iron. Of particular interest is nitrogenase, found in nitrogen fixing bacteria capable of converting atmospheric nitrogen into ammonia. Considerable progress has been made in defining the chemical state of molybdenum in these compounds. There is good evidence that molybdo-enzymes contain two molybdenum atoms per molecule in the oxidation state of + 3 and in some cases +4(1,2,3), the ligating atom to the metal being sulphur (4). Most reactions catalysed by molybdenum have been shown to involve the transfer of two electrons or multiples thereof.

Dialkyldithiocarbamate, $R_2N\text{-}CS_2^-$, a ligand known to stabilise unusual oxidation states of metals, forms complexes ligating through sulphur atoms. Such complexes therefore meet the basic structural requirements expected of a model for molybdo-enzymes (Fig.1). One of the compounds in the present study, namely $Mo_2(dtc)_6$, (dtc=dimethyldithiocarbamate) has been recently shown to be able to rduce pyridine N-oxide to pyridine and nitrate ion to nitric acid and No (5). In view of the fact that electron transfer is an important property essential for the activity of molybdo-enzymes, it was thought pertinent to study the electrical conductivity of dithiocarbamate complexes of Mo and those of other transition metal ions such as Fe^{3+} and Cr^{3+}.

Despite the fact that these complexes were known for a considerable length of time, to our knowledge, no such study has been made until now. In this paper we report our results of conductivity measurements made on pellet samples of these complexes and make an attempt to explain these results in relation to the structure of the compounds.

Experimental:

Dithiocarbamate complexes of molybdenum (III), and molybdenum (IV) were prepared by the method of Brown et al. (5,6), by the oxidative decarbonylation of molybdenumhexacarbonyl, $Mo(CO)_6$, using tetramethylthiuramdisulphide in dry toluene under nitrogen atmosphere. $Mo_2(dtc)_6$ is obtained when the reaction is caried out in the dark, and $Mo(dtc)_4$ is obtained when the reaction is carried out in presence of light. $Fe(dtc)_3$ was prepared using ferric chloride and sodium dimethyldithiocarbamate in aqueous solution. $Cr(dtc)_3$ was prepared by the oxidative decarbonylation of chromium hexacarbonyl using tetramethylthiuramdisulphide in chlorobenzene.

All the samples were analytically pure. They were thoroughly dried and ground to a fine powder. The powder was pressed into pellets of diameter $\sim$ 1.3 cm and thickness 1 - 2 mm in a stainless steel die using 10 KN/cm^2 pressure. Each pellet was made with a layer of graphite powder on either side to act as electrodes. Electrical conductivity of the pellets was measured using a HP 4192-A LCR meter and at TRIO model DL 707 digital multimeter.

Results and Discussion:

The d.c. electrical conductivities of the complexes investigated are given in Table 1.

Table 1

Room temperature d.c. conductivities of dithiocarbamate complexes

		Oxidation state of metal	Electronic config.	Conductivity $(ohm.cm.)^{-1}$
1.	$Mo_2(dtc)_6$	+ 3	$4d^3$	2.5×10^{-3}
2.	$Mo(dtc)_4$	+ 4	$4d^2$	1.7×10^{-3}
3.	$Fe(dtc)_3$	+ 3	$3d^3$	$< 10^{-6}$
4.	$Cr(dtc)_3$	+ 3	$3d^5$	$< 10^{-6}$

Molybdenumdithiocarbamate 1 and 2 have conductivities of the order $10^{-3}(\text{ohm.cm})^{-1}$ at room temperature (25°C) while the compounds 3 and 4 respectively of Fe(+3) and Cr(+3) are poorly conducting. Although definitive X-ray evidence for the structure is lacking, molybdenum compound 1, $Mo_2(dtc)_2$, is believed to be a dimer in the solid state (5,7,8). A partial structure determination and the reported low magnetic moment (0.7 - 0.8 B.M.) for this complex (5,8) suggest appreciable Mo-Mo interactions. Brown et.al have proposed a dimeric structure for this complex, where the co-ordination geometry around each Mo(+3) ion is described as monocapped trigonal prismatic (5). On this basis the ground state electronic configuration is $(d_{x^2-y^2})^2(d_{yz})^1$ and weak metal-metal bonding could result from overlap of the d_{yz} orbitals of the two metal centers.

Although the exact positions of the molecular orbitals from such metal-metal interactions is difficult to estimate, the observed electrical conductivity could arise from d-d transitions from a predominantly d_{yz} orbital to either d_{xz} or d_{z^2}. This assignment is in agreement with the observation of Mitchell (8) that low intensity peaks are observed in the near IR region for Mo(III) complex 1, in its reflectance spectrum. These absorption peaks have been assigned to d-d transitions by Mitchell. According to the estimate of the energy band gap obtained after our recent studies of the variation of the condutivity with temperature, this compound is expected to have an absorption edge in the vicinity of 0.6 eV in the near IR region.

Compound 2, having Mo(IV) with d^2 electronic configuration is diamagnetic. Structure elucidation by X-ray analysis has shown that the MoS_8 co-ordination geometry is that of slightly distorted triangular dedecahedron (9). The diamagnetism in this compound could be explained using molecular orbital and crystal field approaches.

The situation in respect of $Fe(dtc)_3$ and $Cr(dtc)_3$ is less complicated; $Fe(dtc)_3$ has triagonal prismatic geometry with D_3 symmetry whereas $Cr(dtc)_3$ has an almost regular octahedral symmetry. Both of these compounds have the theoretically predicted spin-only magnetic moments, although in the iron complex the temperature dependent $^6A_1$ - $^2T_2$ electron spin cross-over is well known to exist. There is no indication of metal-metal interactions in these complexes.

Ligand fields in the Mo(III) and Mo(IV) complexes provide closely spaced non degenerate d-energy levels and hence low energy electronic transitions. Furthermore, owing to the realatively more diffused nature of the 4d orbitals in comparison to 3d orbitals, there could be weak interactions among metal centers of these molecules in the solid state giving rise to energy bands involving d levels referred to earlier. This could explain the observed conductivity Mo(III) and Mo(IV) complexes.

In $Fe(dtc)_3$ and $Cr(dtc)_3$, which have more symmetrical ligand fields, d-level separation is large. This, coupled with the less spatial extension of 3 d orbitals, makes these compounds poor conductors of electricity.

References:

1. E.J. Stiefel, Prog. Inorg. Chem., 22, 1 (1977).
2. K.B. Swedo and J.H. Enemark, J. Chem. Ed., 56, 70 (1979).
3 W.H. Orme-Johnson, G. Jacob, M. Henzel and B.A., Averill, Adv. in Chem. Series, 162, 389 (1976).
4. K.O. Hodgson et.al., J.Am.Chem. Soc., 100, 3398, 3814 (1978).
5. D.A. Brown, W.K. Glass and K.S. Jasim, Int. Conf. on Molybdenum and molybdenum compounds, Toronto, 156 (1982).
6. D.A. Brown, B.J. Gordon, W.K. Glass and C.J. O'Daly, Proc. 14th Int. Conf. Co-ordination Chem., Toronto, 1972, p.646.
7. R. Lancashire and Thomas D. Smith, J. Chem. Soc. (Dalton) 845 (1982).
8. Philip, C.H. Mitchell and R.D. Scarle, J. Chem. Soc. (Dalton), 110 (1975).
9. J. Willemse, J.A. Cars, J.J. Steggerda and C.P. Keijzers in Structure and Bonding , New York, Springer, 1976, Vol.28, p.83.

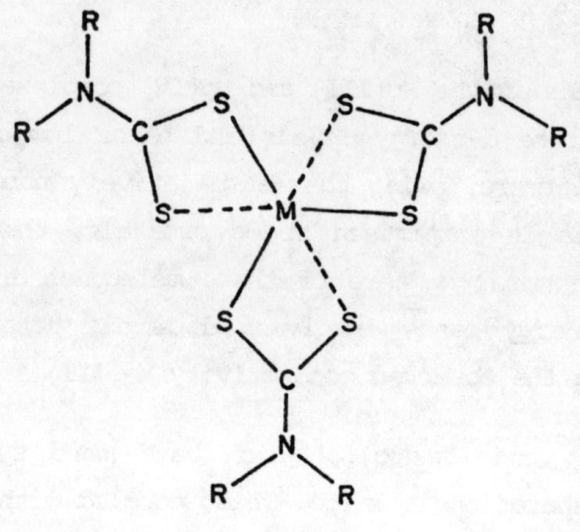

(a)

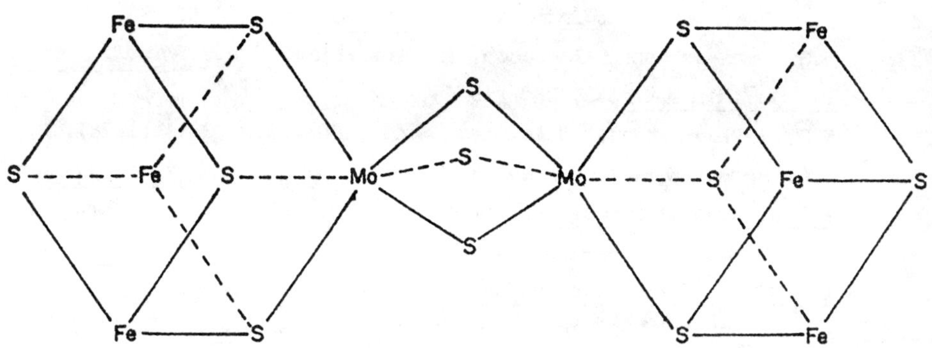

(b)

Figure 1. Structural similarity between metal - dithiocarbamate complexes (a) and molybdo - enzymes (b)

PHASE TRANSITION SHIFT OF CORE LEVEL BINDING ENERGIES OF SOME GROUP I A AND II A METALS

J. S. H. Quintus Perera
Department of Chemistry, University of Peradeniya
Peradeniya, Sri Lanka

1. Introduction

For many years there has been a tacit assumption that core electron binding energies of elements were affected by chemical changes, but not by physical changes [1].

In the past, the free atom core binding energies of metals were estimated by combining the X-ray fluorescence data observed for the solid and optical data for free atoms. This estimation assumes that X-ray fluorescence frequencies are independent of the physical state of the element and the values thus calculated were invariably 3-5 eV for more higher than the corresponding binding energies directly measured for the condensed phase even after the work function correction. It was first thought that this observed difference was due to the invalidity of the assumptions used to estimate the free atom binding energies.

With the development of gas phase X-ray photoelectron spectrometers with high-temperature sample chambers, direct determination of core level binding energies of some metal in their free-atomic state was possible. These values also differ from the corresponding condensed state binding energy values by approximately the same amount, confirming that this difference was in fact due to phase transition.

This phase-transition shift, $E_b(k, M, A)$ can be written as,

$$E_b(k, M, A) = E_b^v(k, A) - E_b^v(k, M) \tag{1}$$

where $E_b^v(k, A)$ and $E_b^v(k, M)$ are the vacuum referenced core binding energies for an electron in the k-th level for the free atom and the metallic condensed state, respectively.

2. Experimental

A high-temperature X-ray photoelectron spectrometer with a high-temperature sample handling facility was constructed in the University of British Columbia and the binding energies of selected core levels of Na, K, Rb, Cross-section, Mg, Ca, Sr, and Ba atoms in the gas phase were determined using this spectrometer. The corresponding phase-transition shifts of these elements were calculated using Equation (1) and these values are given in Table 1.

3. Results and Discussion

Photoionization of a core level in an atom leads to relaxation of the remaining passive electrons. The net result of this is stabilization of the final state, which leads to the definition of the quantity termed relaxation energy. This relaxation energy varies from free atoms to molecules to solids. If a core level of a metal in its condensed state is ionized, the total relaxation energy following the core ionization can be written as the sum of two terms:

$$E_{\text{relaxation}} = E_{\text{intra-atomic}} + E_{\text{extra-atomic}} \tag{2}$$

$E_{\text{intra-atomic}}$ is more or less the same as the relaxation energy for the free atom. $E_{\text{extra-atomic}}$ is an additional relaxation energy due to the relaxation of the valence (conductance) band electrons in the condensed state.

To theoretically estimate the extra-atomic relaxation energy term in metals Shirley et al. [1, 2] used a theoretical model based on the assumption that extra-atomic relaxation occurs through screening of the hole state by the formation of a semi-localized exciton. They further assumed that the exciton state has the symmetry of the lowest unbound state in the conduction band and that the exciton wave function is found only in the neighbor-

Table 1
Experimental phase transition shifts of group I A and II A metals determined in this work and some theoretical values

Element	Level	Phase-Transition Shift eV	
		Experimental	Theoretical
Na	1s	5.1	5.3[a]
K	$2p_{1/2}$	3.6	3.0[b]
K	$2p_{3/2}$	3.6	3.0[b]
Rb	$3p_{1/2}$	3.4	
Rb	$3p_{3/2}$	3.4	
Cs	$3d_{3/2}$	3.6	
Cs	$3d_{5/2}$	3.6	
Mg	1s	4.8	5.1[c]
Ca	2s	6.5	7.0[b]
Ca	$2p_{1/2}$	6.0	7.0[b]
Ca	$2p_{3/2}$	6.0	7.0[b]
Sr	3d	5.6	
Ba	$3d_{3/2}$	5.0	
Ba	$3d_{5/2}$	5.0	
Ba	$4d_{3/2}$	5.9	
Ba	$4d_{5/2}$	5.9	

Notes: a) Ref. [4], b) Ref. [5], c) Ref. [6].

hood of the hole state. Using this model they further argued that the extra-atomic relaxation term is by far the largest contributor to the phase transition shift in metals. Some theoretical values of phase-transition shift calculated by Shirley et al. [4, 5, 6] using their model are listed in Table 1.

These results show, although the semi-localized exciton model seems to over estimate the extra-atomic relaxation, the trends of binding energy shifts predicted by this model are in excellent agreement with the experiment. In addition, Siegbahn et al. and Citrin et al. [3] have determined the core level binding energies of rare gas atoms embedded in metal foils and these values were also found to be 2-5 eV lower than those reported for free atoms. As the interaction between the gas atom and the host metal atoms is more of a physical nature, the shift may be indicative of the degree of extra-atomic relaxation in the metallic state.

In summary, the phase-transition shift due to core ionization is believed to be caused by a number of facts, such as:

a. screening of the vacant orbital via extra-atomic relaxation of metal valence electrons;

b. changes in the repulsive potential experienced by the core electrons in metals; and

c. changes in the electronic configuration when going from solid to vapor, each contributing to a different degree in different metals. Although (c) is not relevant in the present context, it is important in transition metals where the common configuration in the vapor phase is $nd^x(n + 1)s^2$ as opposed to the $nd^{x+1}(n + 1)s^1$ configuration that represents the solid. Chromium which exhibits the lowest phase-transition shift in the 3d transition series has the same ground state electronic configuration in both the free atom and in the condensed phase (viz. $3d^5\ 4s^1$).

The importance of phase-transition shift observed for metals is now well recognized. For example, the phase-transition shift may prove useful in studying the nature of adsorbate-adsorbent interactions.

References

1. Shirley, D. A., *Chem. Phys. Lett.*, vol. 16, p. 220, 1972.

2. Ley, L., Kowaczyk, S. P., McFeely, F. R., Pollak, R. A., Shirley, D. A., *Phys. Rev. B*, vol. 8, p. 2392, 1973.
3. Citrin, P. H., Hamann, D. R., Hufner, S., *Buss. Am. Phys. Soc.*, vol. 1, p. 55, 1973.
4. Kowalczyk, S. P., Ley, L., McFeely, F. R., Pollak, R. A., Shirley, D. A., *Phys. Rev. B*, vol. 8, p. 3583, 1973.
5. Shirley, D. A., Martin, R. L., Kowalczyk, S. P., McFeely, F. R., Ley, L., *Phys. Rev. B*, vol. 15, p. 544, 1977.
6. Ley, L., McFeely, F. R., Kowalczyk, S. P., Jenkins, J. G., Shirley, D. A., *Phys. Rev. B*, vol. 11, p. 600, 1975.

POLARIZATION AND DISCHARGE CHARACTERISTICS OF THIN FILM SOLID STATE CELLS WITH Cu^+ ION CONDUCTORS

M. A. K. L. Dissanayake, J. Karunamuni
Department of Physics, University of Peradeniya
Peradeniya, Sri Lanka

Abstract: Thin solid films of CuI-CuCl mixed phase electrolyte has been formed on copper substrate by an electrolytic deposition technique. All solid state thin film cells of configuration Cu/CuI: CuCl/Mg have been fabricated and their polarization and discharge characteristics have been investigated. The open circuit voltage of a typical cell was found to be ~1.1 V. The capacity and energy density of the cell were ~1.5 mAh and ~100 mWh/cm^3.

Introduction:

Solid state electrochemical cells have many advantages over conventional aqueous cells due to absence of liquid leakage or gasing, light weight, rugged structure and the possibilities of making thin film microbatteries and operation over a wide temperature range. Many of the solid state cells that are being used or being developped at present are based on either Ag^+ ion conductors or Li^+ ion conductors which are expensive or chemically unstable[1]. Copper ion conducting solid electrolytes have the advantage of low cost and the cells based on these electrolytes have the possibility of using copper metal as an electrode material. It is,therefore, interesting to study characteristics of Cu^+ ion based solid state cells. Recent work on several promising solid copper ion conductors and cells based on these materials can be found in references 2,3. In this paper, we report the polarisation and discharge characteristics of thin film solid state cells using CuI: CuCl mixed phase as the electrolyte and Cu and Mg as electrodes.

Experimental:

Thin films of CuI:CuCl mixed phase solid electrolyte was prepared on copper substrate by electrodeposition method described in reference (4). 100 ml of NH_4Cl solution was made up with A.R.reagent and distilled water and the electrolyte solution was compounded by adding six drops of HI solution of concentration 45 % into the 100 ml of 0.1 M NH_4Cl solution. Two perfectly cleaned copper foils were mounted with their surfaces parallel to each other and at a fixed separation of about 5 cm. and were dipped into the vessel containing the electrolyte. The electrolytic current was selected to be $\sim$ 5 mA/cm^2. During electrolysis, the anions of Cl^- and I^- would migrate towards the anode and react with Cu^+ ions supplied by the anode to form a mixed polycrystalline film of CuI:CuCl which would be depositted on the anode surface. The thickness of the film would depend on the duration of the electrolysis. The thickness of the film after a deposition time of five minutes should be about 100 μm. After the electrylysis, the copper substrate with the CuI:CuCl film on it was taken out of the electrolytic bath and washed with ethanol and diethyl ether and dried in a dessicator. Cells were fabricated by mechanically pressing a thin Mg foil against the electrolyte film . Currents and voltages were measured using a TRIO digital multimeter with 0.1 μA and 0.1 mV sensitivities.

Results and Discussion:

The OCV of a typical cell of configuration Cu/CuI:CuCl/Mg was ~ 1.1 V. Fig. 1 shows the room temperature discharge curves for 1MΩ and 100 kΩ load resistors. The average current during the discharge was about 2 μA and 12 μA respectively. Based on these values the capacity of the cell is estimated to be ~1.5 mAh assuming a discharge time of 120 hours. For a cell of film thickness 100 μm, the energy density is estimated to be ~100 mWh. Fig. 2 shows a set of polarisation curves taken after different times of discharge through the 100 kΩ load resistor. It can be seen that the typical cell behaviour is maintained even after 4 days of discharge at 12 μA.

The internal resistance of the cell was ~7 MΩ and was increasing with the time of discharge. This high value could be attributed to the very low ionic conductivity of CuI and CuCl at room temperature. The reported values are 10^{-9} (ohm cm $)^{-1}$ and 10^{-12}(ohm cm$)^{-1}$ respectively[5]. A white reaction product appears at the electrolyte/Mg interface after several hours of discharge. The increase in internal resistance is possibly caused by this layer.

During the discharge process of the Cu(+)/CuI:CuCl/Mg(-) cell, the charge transfer reactions at the respective electrodes can be represented as follows:

At the Mg anode, oxidation takes place via the reaction,

$$Mg \longrightarrow Mg^{++} + 2e$$

The Mg^{++} ions react with the electrolyte to form MgI_2+$MgCl_2$ mixed product, white in colour, at the electrolyte/Mg interface.

$$2\ Mg^{++} + 2\ CuI + 2\ CuCl \longrightarrow MgI_2 + MgCl_2 + 4\ Cu^{+}$$

The Cu^+ ions would migrate through the electrolyte to the Cu cathode. In addition to the above reaction, a thin layer of metallic copper is also formed at the electrclyte/Mg interface. This should be due to the following reactions:

$$CuI + Mg \longrightarrow MgI_2 + Cu$$

$$CuCl + Mg \longrightarrow MgCl_2 + Cu$$

The charge transfer reaction at the copper cathode, where the reduction takes place, can be represented as:

$$Cu^+ + e \longrightarrow Cu$$

As a result, metallic copper is depositted on the copper cathode surface. Therefore, the overall cell reaction during the discharge process, would gives rise to a depletion of the CuI:CuCl electrolyte, deposition of metallic copper at the two interfaces and the formation of MgI_2+$MgCl_2$ layer at the electrolyte/Mg interface.

Conclusion:

Polarisation and discharge characteristics of solid state cells with CuI:CuCl mixed phase thin film electrolyte and copper and magnesium electrodes have been studied. The open circuit voltage of the cell was ~ 1.1 V. The estimated capacity and the energy density are ~1.5 mAh and ~ 100 mWh/cm^3 respectively. The cell, however, has a rather high internal resistance which increases with discharge time due to the formation of reaction products at the electrolyte/Mg interface.

References:

1. Modern Batteries, Collin A.Vincent, Edward Arnold (1984).
2. O.Yamamoto in "Materials for Solid State Batteries", ed. B.V.R.Chowdari and S.Radhakrishna, World Scientific Co. ,Singapore (1986) 263.
3. O.Yamamoto, Y.Takeda and R.Kanno, ibid 275.
4. Yu Wen-hai, Zhu Bin and Bheng Qing, ibid 481.
5 J.B.Wagner and C.Wagner, J.Chem. Phys. 26 (1957) 1597.

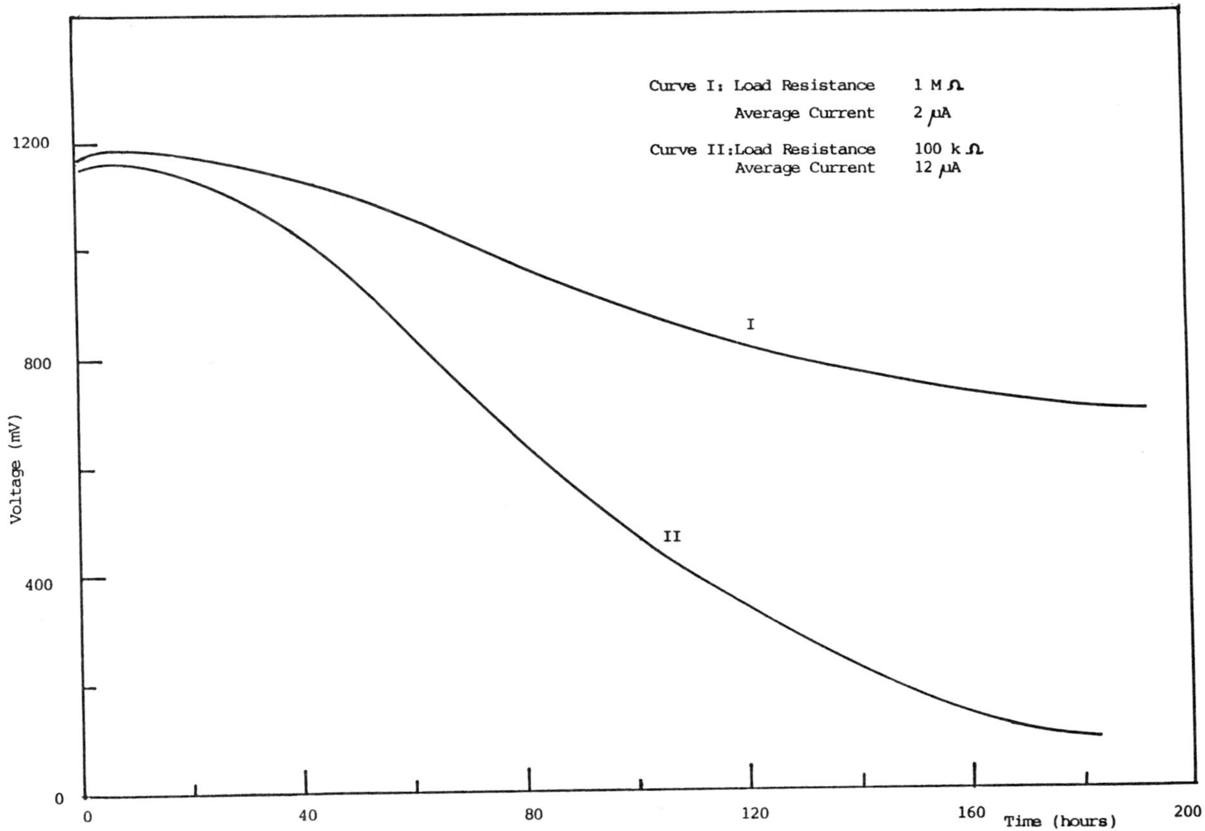

Fig.1: Discharge curves for the cell Cu/CuI:CuCl/Mg at constant load.

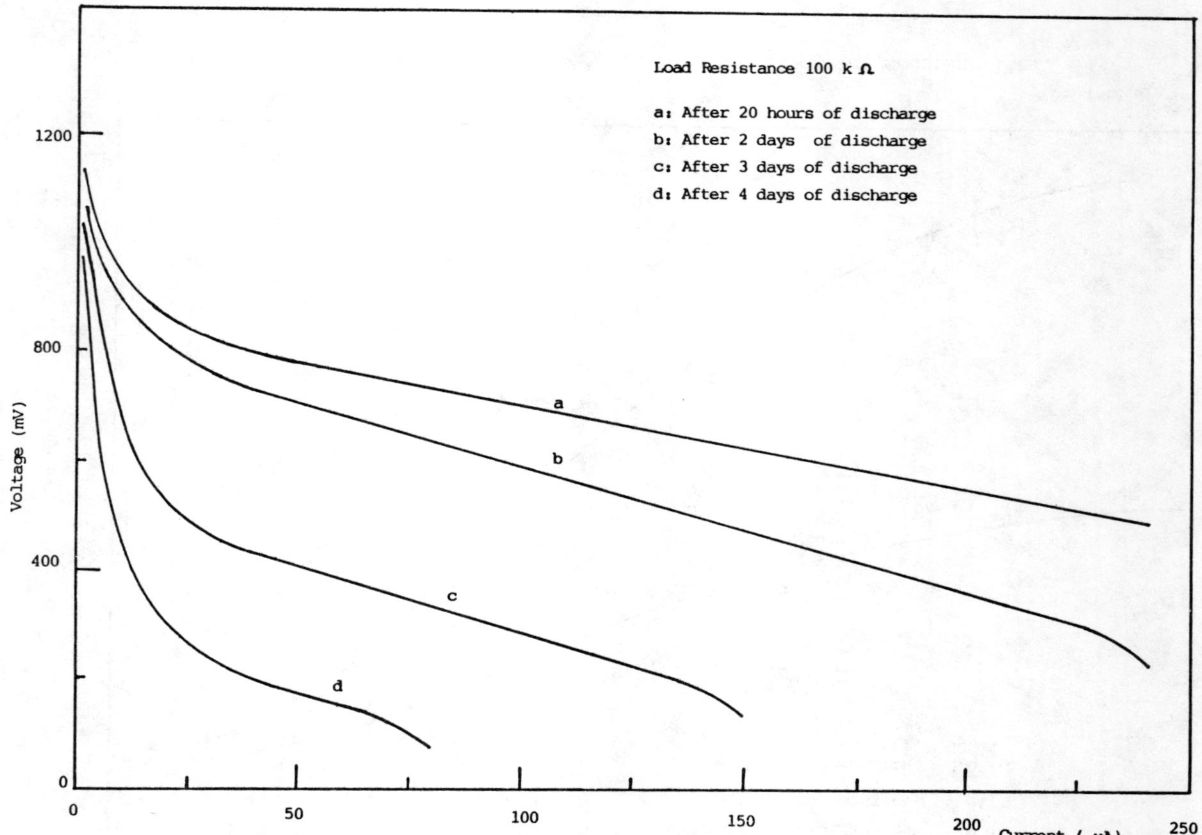

Fig.2: Polarisation curves for the cell Cu/CuI:CuCl/Mg.

MICROCOMPUTER BASED COMPLEX IMPEDANCE MEASUREMENTS ON SOLID STATE IONIC MATERIALS

B. V. R. Chowdari, K. Radhakrishnan
Department of Physics, National University of Singapore
Kent Ridge, Singapore 0511

Abstract: The performance of an automated complex impedance measuring system which was recently assembled in our laboratory for electrical characterization of ionic conductors is discussed. Procedures for measurements, data acquisition and analysis are described. Simulated complex impedance and modulus spectra using standard resistor/capacitor combinations are presented and the results obtained for some of the real solid state ionic materials using this automated system are given. Emphasis is placed on the mixed cationic effect in the Li/Ag-B_2O_3-P_2O_2 system and the results are compared with materials exhibiting similar phenomena.

ELECTRICAL CONDUCTIVITY OF Li_2-SO_4-$MgSO_4$ EUTECTIC SYSTEM

M. A. Careem, M. A. K. L. Dissanayake
Department of Physics, University of Peradeniya
Peradeniya, Sri Lanka

Abstract: **Electrical conductivity of several compositions of the Li_2SO_4-$MgSO_4$ system has been measured up to 550°C. A maximum in the conductivity has been observed for the eutectic mixture with about 22 mol-% $MgSO_4$ composition. The conductivity enhancement has been attributed to the increased interfacial conduction in the finely dispersed grains in the two-phase eutectic mixture.**

1. Introduction

Pure Li_2SO_4 undergoes a steady-solid phase transition at 577°C from a low-temperature β-phase with a monoclinic structure to the high-temperature α-phase with an f.c.c. structure. α-Li_2SO_4 is a solid electrolyte having a very high ionic conductivity of ~3 ohm^{-1} cm^{-1} close to its melting point at 860°C [1]. Various binary and ternary systems with α-Li_2SO_4 as one component have been studied in the past [2-5]. β-Li_2SO_4 is a poor ionic conductor with a conductivity of ~10^{-11} ohm^{-1} cm^{-1} at room temperature. The ionic conductivity of such poor conductors can, however, be increased considerably by dispersing a relatively low conducting second phase to form a heterogeneous system. Such "composite electrolytes" exhibit enhanced ionic conductivity compared to the individual phases. The first such example was the enhancement of Li^+ ion conductivity in the LiI-Al_2O_3 system reported by Liang in 1973 [6]. Since then a variety of composite electrolytes have been reported in the literature [7]. Li_2SO_4-Al_2O_3 system [8] and Li_2SO_4-Li_2CO_3 system [9] are among the few composite electrolytes with β-Li_2SO_4 as one phase where enhancement of Li^+ ion conductivity has been observed. Similar enhancement in conductivity with a maximum conductivity for the eutectic mixture has been reported by us for the Li_2SO_4 $CaSO_4$ two phase system [10].

The phase diagram of the Li_2SO_4-$MgSO_4$ system as determined by Ljungmark has features similar to that of Li_2SO_4-$CaSO_4$ system [2]. At temperatures below 550°C, a two-phase region extends up to about 65 mol-% $MgSO_4$ composition and the eutectic mixture corresponds to about 22 mol-% $MgSO_4$ composition. Therefore in this system a conductivity enhancement can be expected at the eutectic mixture, similar to the one observed for the Li_2SO_4-$CaSO_4$ system. In this paper we report some preliminary conductivity measurements in the Li_2-SO_4-$MgSO_4$ two-phase system.

2. Experimental

Reagent grade Li_2SO_4 and $MgSO_4$ purchased from Merk, Germany, were used as the starting materials. The salts were first dried in air at 400°C for ~48 hours. The different compositions were thoroughly mixed and ground and heated to melt in alumina crucibles. After solidification the mixtures were ground to a fine powder. Pellets of about 2 mm thickness and 13 mm diameter were pressed in a stainless steel die using a pressure of 300 M Pa. The pressed pellets were annealed at 475°C in air for 24 hours before evaporating gold electrodes onto them. For electrical conductivity measurements, each pellet was inserted between two silver spiral electrodes and the pellet assembly was placed in the middle of a Heraus tube furnace. Conductivity was determined using a.c. impedance measurements taken at 25°C intervals during the cooling cycle starting at 550°C. The complex impedances from 10 Hz to 10 M Hz were measured using a Hewlett-Packard 4192 A LCR meter with a 20 mV signal applied to the sample. The furnace temperature was controlled by an Eurotherm 810 temperature controller, and the sample temperature was measured using a Chromel-alumel thermocouple placed very close to the pellet.

3. Results and Discussion

The electrical conductivity at various temperatures for four different compositions of the Li_2SO_4-$MgSO_4$ system plotted in the form of log (σT) vs 1/T are shown in Figure 1. The data for pure β-Li_2SO_4 were taken from reference [9]. The conductivity variations with compositions at 375°C, 450°C, 500°C are shown in Figure 2. These graphs clearly show that at temperatures below 550°C the conductivity first increases with increasing $MgSO_4$ content up to the eutectic composition and then decreases with further addition of $MgSO_4$. At 500°C the conductivity of the eutectic mixture is $\sim 3.6 \cdot 10^{-3}$ ohm^{-1} cm^{-1} whereas that of pure β-Li_2SO_4 is $\sim 9.5 \cdot 10^{-5}$ ohm^{-1} cm^{-1}. This corresponds to a conductivity enhancement of ~35 times. The conductivity enhancement observed in the Li_2SO_4-$MgSO_4$ system is similar to that observed for composite electrolyte systems such as LiI-Al_2O_3 [6], AgI-Al_2O_3 [11], Li_2SO_4-Al_2O_3 [8], Li_2SO_4-Li_2CO_3 [9], Li_2SO_4-$CaSO_4$ [10], and AgI-AgBr [12]. In these systems conductivity enhancement has been attributed to increased conduction in the interface region between grains. According to the phase diagram of the Li_2SO_4-$MgSO_4$ system, at temperatures below 550°C, the Li_2SO_4-$MgSO_4$ system exists as a two phase mixture. Therefore the interface between grains in the two-phase region is expected to provide a favorable conduction path to the mobile ions. In the case of the Li_2SO_4-Li_2CO_3 system, an SEM study has confirmed the finely dispersed character of the grains in the eutectic mixture [9]. In the two-phase region of the Li_2SO_4-$MgSO_4$ system the eutectic mixture is very likely to have the minimum crystal grain size giving rise to the maximum interfacial area between grains. Thus the interfacial contribution to the overall ionic conductivity is expected to be highest in the eutectic mixture. Therefore the eutectic mixture should exhibit the maximum overall conductivity at a given temperature.

Acknowledgement

We would like to thank C. N. Wijesekara for assisting in some of the conductivity measurements. We are also indebted to the International Seminars of Uppsala University, Sweden for providing equipment and training, and to the University of Peradeniya for financial assistance.

References

1. Kvisit, A., Lunden, L., *Z. Naturforsch*, vol. 20a, p. 235, 1965.
2. Ljunmark, H., PhD Thesis, University of Gothenburg, 1974, unpublished.
3. Heed, B., Lunden, A., Schroeder, K., In: *Relativity of Solids*, Ed. by J. Wood, O. Lindquist, C. Helgesson, and N. G. Vannerberg, New York: Plenum Press, p. 713, 1977.
4. Heed, B., Lunden, A., Schroeder, K., *Electrochem. Acta*, vol. 22, p. 705, 1977.
5. Aronsson, R., Janson, B., Knape, G., Lunden, A., Nilsson, L., Sjoblom, C. A., Torell, L., *J. Physique Colloq.*, vol. 41, pp. 6-35, 1980.
6. Liang, C. C., *J. Electrochem. Soc.*, vol. 120, p. 1289, 1973.
7. Poulsen, F. W., In: *Transport Structure Relations in Fast Ion and Mixed Conductors*, Ed. by F. W. Poulson, N. H. Anderson, K. Clausen, S. Skaarup, and O. Toft Sorenson, Roskilde: Riso National Laboratory, p. 67, 1985.
8. Liquan, Chen, Zongyuon, Zhao, Chaoyang, Wang, Zirong, Li., *Acta Physica Sinica*, vol. 34, p. 1027, 1985.
9. Dissanayake, M. A. K. L., Mellander, B. E., *Solid State Ionics*, vol. 21, p. 279, 1986.
10. Careem, M. A., Dissanayake, M. A. K. L., In: *Materials for Solid State Batteries*, Ed. by B. V. R. Chowdari and S. Radhakrishna, Singapore: World Scientific Publ. Co., pp. 298, 1986.
11. Chowdhary, P., Tara, V. B., Wagner, J. B., Jr., *J. Electrochem. Soc.*, vol. 132, p. 123, 1985.
12. Shahi, K., Wagner, J. B., Jr., *Solid State Ionics*, vol. 12, p. 511, 1984.

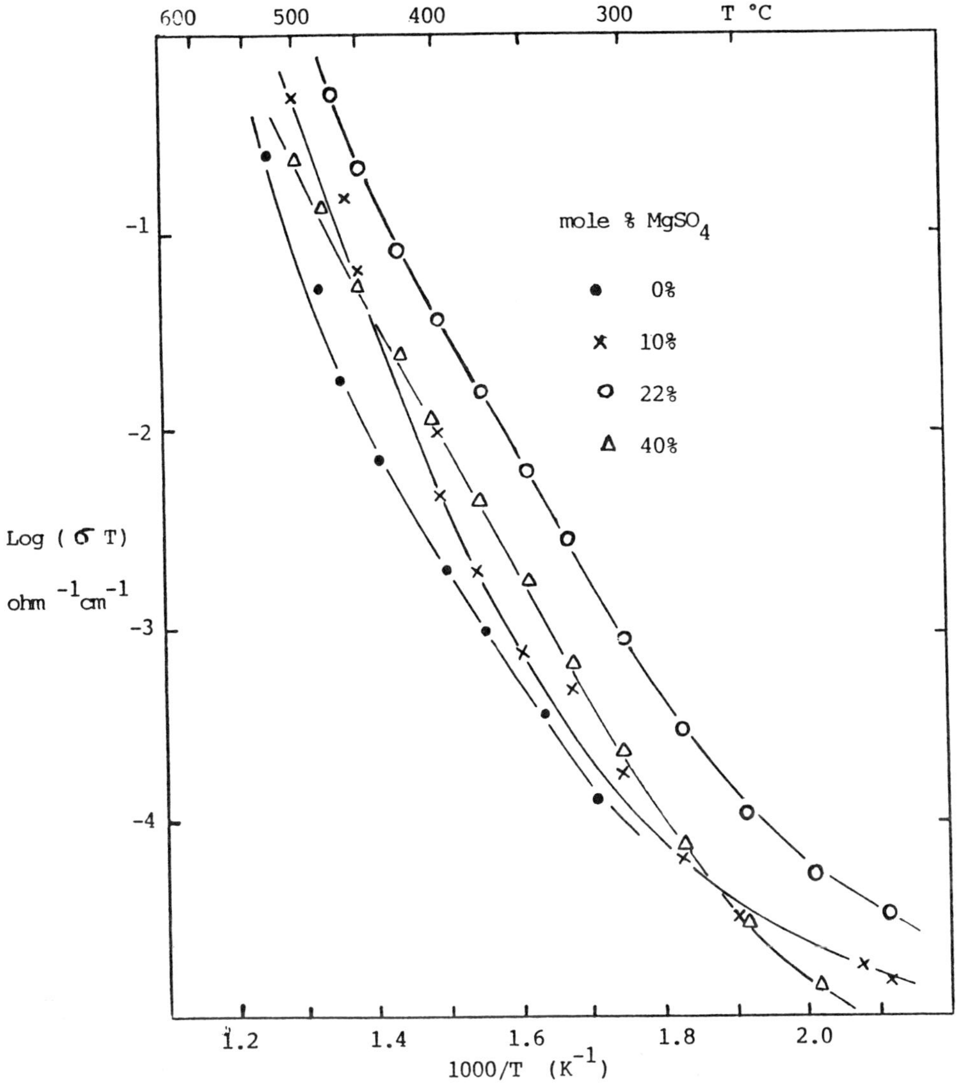

Fig. 1 Log(σT) vs 1/T plots of Pure Li_2SO_4 and of mixtures upto 40 mole % $MgSO_4$.

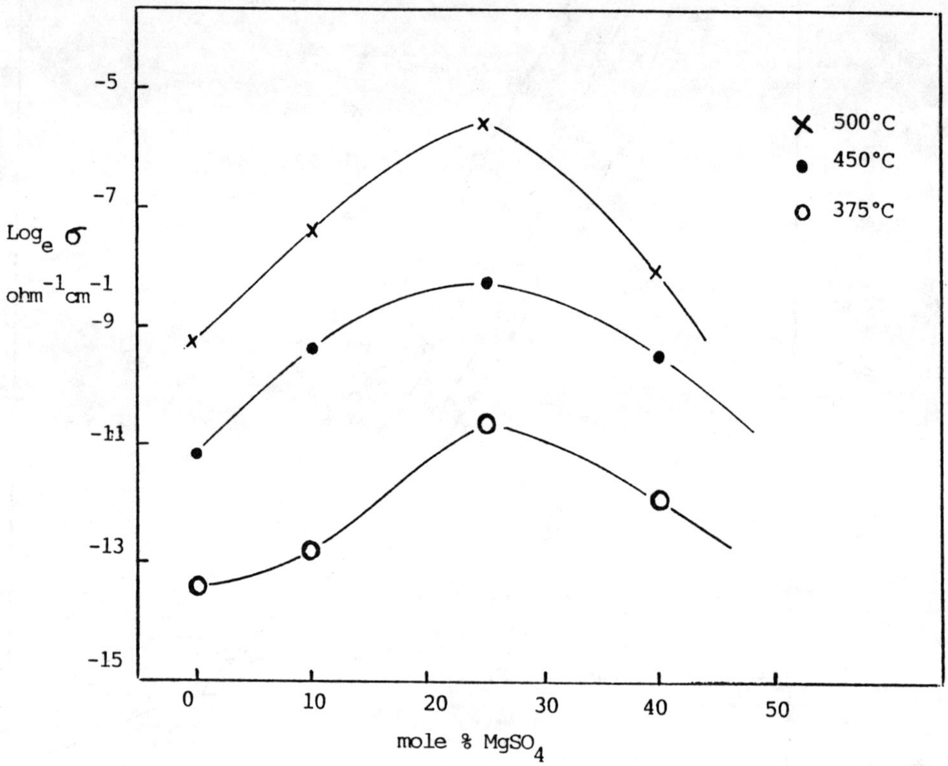

Fig. 2 Variation of electrical conductivity with mole % $MgSO_4$ in Li_2SO_4 for three different temperatures.

DISCHARGE CHARACTERISTICS OF SOLID STATE CELLS WITH CuI-Al_2O_3 ELECTRODES

P. A. G. D. Dharmasena, O. A. Ileperuma
Department of Chemistry University of Peradeniya
Peradeniya, Sri Lanka

M. A. K. L. Dissanayake
Department of Physics, University of Peradeniya
Peradeniya, Sri Lanka

Abstract: Solid state cells with CuI + xAl_2O_3 as the electrolyte and Cu and Mg as the electrode have been fabricated. Maximum open-circuit voltage of ~1.0 V and minimum internal resistance have been obtained for cells with 30-35 mol-% Al_2O_3 content. Polarization and discharge characteristics for different loads have been studied. The capacity and the energy density of the cells have been estimated to be 10 mWh and 100 mWh/cm^3, respectively.

Introduction

All - solid state cells have several advantages such as the absence of any possible liquid leakage or gasing, possibility of miniaturization and operation over a wide temperature range, rugged structure and long storage life. Although AgI based solid electrolytes such as $Rb\,Ag_4\,I_5$[1,2] with high Ag^+ ion conductivity were known from 1960 s, their practical applications in batteries were limited because of the high cost of silver compounds. Copper ion conducting solid electrolytes are promising materials for solid state cells because of the lower cost. A large number of Cu^+ ion conductors with high conductivity based on CuI, CuBr and CuCl have been reported[3,4] by Takahashi et. al. Ionic conductivity enhancement of several orders of magnitude in LiI due to the addition of fine Al_2O_3 particles was first observed by Liang[5] in 1973. Similar effects have been observed later in many other composite electrolyte systems. In the present study we have observed the maximum conductivity enhancement of CuI due to the addition of 30-35 mole % Al_2O_3. All solid state cells with CuI + Al_2O_3 composite electrolyte and Cu - Mg bimetal electrodes have been fabricated and their discharge characteristics investigated.

Experimental

CuI powder was prepared by the reaction of KI with $CuSO_4$ in aqueous solution under a steady stream of SO_2. The resultant solid was filtered and washed well with SO_2 saturated water and ethand. The CuI powder obtained in this manner and Al_2O_3 powder were dried at 150°C for 12h. The mixtures used were weighed to 0.1 mg accuracy. Compressed pellets of pure CuI and of CuI with 10,20,30,35,40 and 60 mole % Al_2O_3 powder were made using a stainless steel die and a pressure of about 10 kN/cm^2. The diameter of the pellets was about 1.3 cm and thickness was about 1.3 cm and thickness was about 2 mm. Pellets used to fabricate the cells were made with a thin layer of fine copper powder on one side. A mechanically pressed pure Mg sheet formed the anode of each cell. The discharge and the polarization characteristics of the cells with configuration $Mg/CuI + x.Al_2O_3/Cu$ were determined using a Trio model DL 707 digital multimeter and a Kipp and Zonen chart recorder.

Results and Discussion

The open circuit voltages of cells with configuration $Mg/CuI + x.Al_2O_3/Cu$, for various Al_2O_3 contents were measured. It was observed that the cells with 30 and 35 mole % Al_2O_3 had the highest open circuit voltage of about 1.0 V and the minimum internal resistance of about 7M . However, the internal resistance of the cells increased rather quickly during the discharge to values greater than 20 MΩ.

The discharge characteristics of cells are shown in Figures 1 & 2 where for each cell, the cell voltage during a discharge through a constant load resistance is plotted against time. It is clear from the figure that the voltage plateau covers a period of about 3 days and then starts to drop to very small values. The polarization characteristics for a cell with 35 mole % Al_2O_3 at different stages during the discharge process is shown in Figure 2. The high value of the internal resistance may be attributed to the low ionic conducting of CuI at room temperature. By extrapolating the values from reported data[9], the room temperature conductivity of CuI should be of the order of $10^{-9} \Omega^{-1} cm^{-1}$. Although we have not attempted to measure the ionic conductivity of the $CuI+Al_2O_3$ system, the conductivity is expected to be higher than that of pure CuI due to the composite effect as observed for LiI and AgI systems. The cell reactions at the electrodes can be written as follows:

At the anode : $Mg \longrightarrow Mg^{2+} + 2e$

At the cathode: $Cu^{+} + e \longrightarrow Cu$

Accordingly, during the discharge process a layer of MgI_2 can be formed at the interface between the Mg electrode and the electrolyte. This could contribute to the increase in the internal resistance of the cell.

Conclusion

An all solid state cell of the configuration Mg/CuI $+x.Al_2O_3$/Cu has been fabricated. Polarization and discharge characteristics of the cells with CuI and 35 mole % Al_2O_3 have been studied. The capacity and the energy density of the cells have been estimated to be 10 mWh and 100 mWh/cm^3 respectively.

References:

01. B.B. Owens and G. R. Argue, Science, 157, (1967), 303.

02. J.N. Bradley and P. D. Greens, Trans Faraday Soc., 63, (1967), 2516.

03. T. Takahashi, O. Yamamoto, S. Yamada and S. Hayashi, J. Electrochem. Soc, 126, (1979), 1654.

04. O. Yamamoto in "Materials for Solid State Batteries", ed. B.V.R. Chowdari, R.S. Radhakrishna, Singapore, World Sicentific Publ. Co ., (1986), 263.

05. C.C. Liang, J. Electrochem Soc., 120, (1973), 1289.

06. T. Jow and J. Bruce Wagner Jr. J. Electrochem. Soc., 126 (1979) 1963.

07 Collins, A., Vincent, Modern Batteries, London, Edward Anorld, 1984.

08. S. Chandra, Superionic Solids, North Holland, 1981.

09. T. Matsui, and J.B. Wagner, J. Electrochem, Sco., 124, (1977), 610.

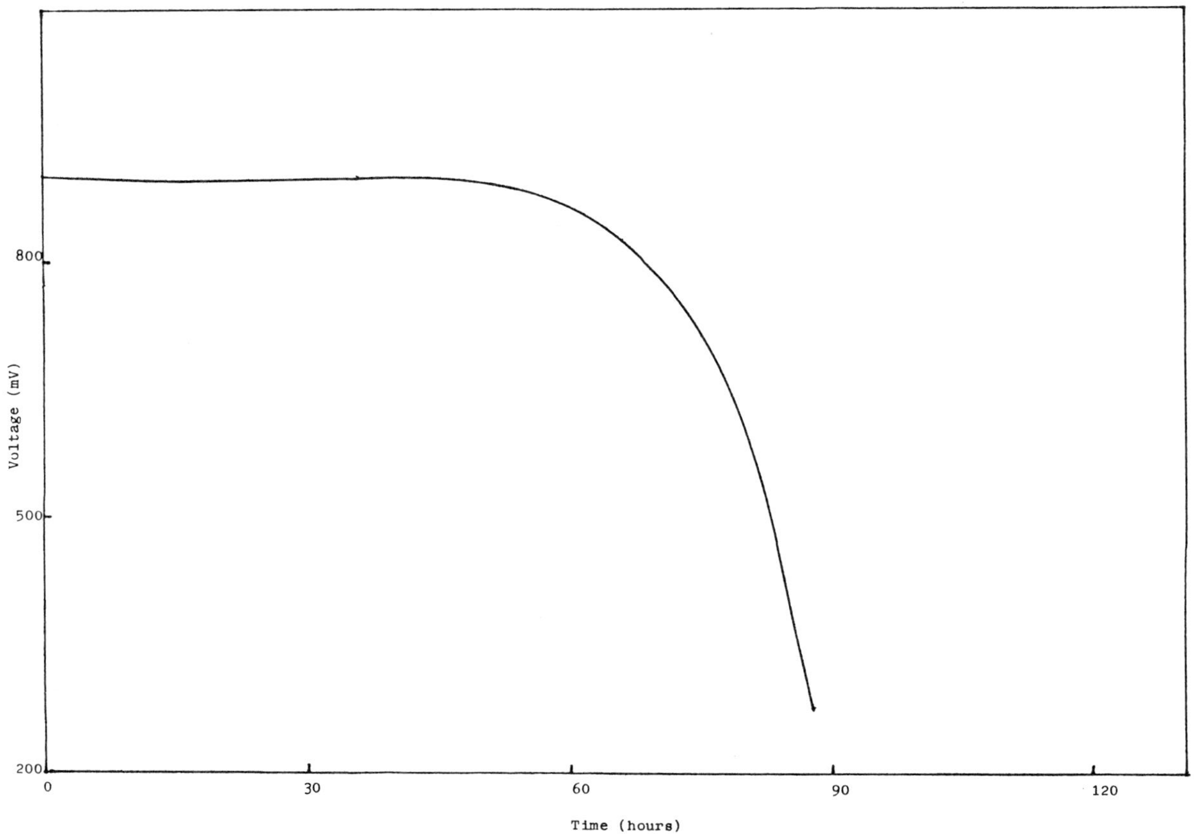

Figure 1 : Curve for cell : Cu/CuI + 35% mole Al_2O_3/mg through 50 K load resistance

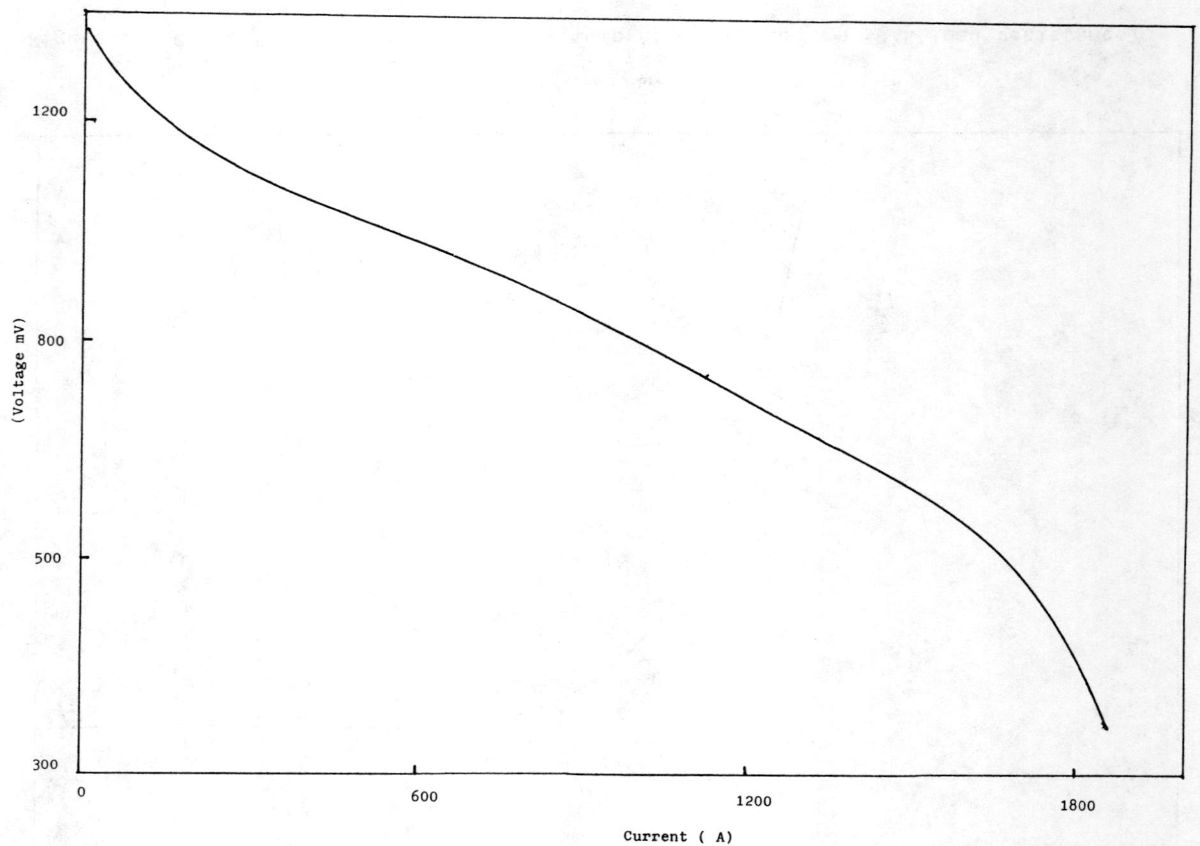

Figure 2: Polarization curve for cell : Cu/CuI + 35% mole Al_2O_3/mg after discharging through 1 kΩ 30 hours.

EFFECT OF QUENCHING TECHNIQUE ON THE ELECTRICAL CONDUCTIVITY OF Li_2O:B_2O_3 SYSTEM

K. Singh, P. R. Gandhi, N. R. Choudhary, S. V. Rokade
Department of Physics, Nagpur University
Nagpur-440010, India

Abstract: **Lithium borate glasses of different compositions have been prepared by using Al-mould, Twin Roller (TR), and Cu-drum quenching techniques and their electrical conductivity has been studied. It is possible to extend the glass formation region up to 70 mol-% Li_2O due to the faster quenching rate in a Cu-drum. The conductivity of amorphous material is more than that of the crystalline due to the increase disorder in the system.**

Introduction:

Solid electrolytes are gaining more importance due to interesting scientific problems and potential applications in the technological devices. Large scale applications are expected for solid electrolytes as separators in batteries. Earlier, this particular area was dominated by crystalline materials (mostly from β-alumina family). Because of the recent progress and practical requirements, much attention has now been focussed on glassy solid electrolytes which have certain advantages over crystalline solids. The ionic conductivity in glasses has isotropic nature and the grain boundary problems do not exist at all. One of the interesting features of glass as solid electrolyte material is the possibility of continuously changing composition. As such any composition can be quenched into a glass if it is cooled sufficiently rapidly to a low enough temperature [1]. The composition parameter is a useful tool for investigating the conduction mechanism in glasses [2]. One of the factors that affect the conductivity is the concentration of the mobile ion species [3]. With this view, Li_2O-B_2O_3 system has been studied. It is possible to obtain increased ionic conductivity for higher alkali ion concentration, provided, the system exhibits glassy nature [4]. The faster quenching technique has made it possible to get glass for a wide composition range.

The extension of glass formation region for Li_2O content upto 65 mol % fast quenching technique (TR) has already been achieved by Ota et al. [5]. It has been observed that glass structure can be manipulated by chaning the thermal history [6]. Thus, the formation of glass is purely dependent on the rate of quenching, because the physical properties such as density, specific volume and viscosity get considerably affected by the method of glass formation [7]. Also the change in quenching rate causes the change in transition temperature T_g, which in-turn, changes the structure and hence the properties [8]. Ito et al have reported the glass formation region upto 80 mol % for Li_2O-B_2O_3 by evaporation technique [9], and explained the variation of electrical conductivity in the light of structure determined by IR and Laser Raman Sepectroscopy. It was therefore, thought imperative to investigate the effect of quenching techninique on the electrical conductivity.

Thus in the present investigation, the compostions 40:60, 50:50, 60:40, 70:30 and 75:25 mol % of $Li_2O-B_p2O_3$ have been prepared by aluminium mould (Al-mould), Twin roller (TR) and copper-drum (Cu-drum) quenching techniques (QT).

Experimental:

B_2O_3 A.R. (99% pure) as supplied by SISCO (India) and Li_2CO_3 (99.9% pure, Moscow, USSR) were used as the starting materials. Five batches with 40:60, 50:50, 60:40, 70:30 and 75:25 mol % $Li_2O:B_2O_3$ were prepared. About 12 gms of the raw material of each batch was weighed and mixed thoroughly under acetone. The dried mixture was taken in porcelain crucible and heated at moderate rate in an electric furnace. The temperature was raised upto 1123 K. The melt was kept at this temperature for 2 hours for homogenization. The glasses were casted by quenching the melt at room temperature in Al-mould, TR and Cu-drum, within glass formation range, the glasses were found to be colourless and transparent. The prepared glasses were characterized by XRD and DTA techniques.

The samples obtained from Al-mould and Cu-drum techniques were polished flat whereas, TR provided the samples in the form of very thin flexes unable to withstand the spring loading tension of the sample holder. So these flexes were finely crushed and pelletised at 10 ton/cm^2. To attain mechanical strength of these pellets, they were sintered at 200°C for 2 hrs. A good ohmic contact was achieved by the application of silver paint to the flat surfaces of the samples, followed by baking at 100°C.

The fixed frequency (1KHz) measurements were carried out to have a quick scanning of the conductivity range of the various samples. For impedance analysis, the samples which showed an encouraging magnitude of conductivity were considered. A.C. electrical conductivity was measured as a function of frequency (5 Hz-13MHz at various temperatures in the range 463-623 K with Hewlett Packard 4192A LF impedance analyser.

Result and Discussion:

To study the effect of different quenching techniques on the conductivity of Li_2O-B_2O_3 system, XRD and DTA data were carefully analysed. The crustalline phases along with the compostion for three different QT have been tabulated in Table I. This table reveals that in case of Al-mould and TR, the samples above 40:60 mol % crystallized, where, as in Cu-drum upto 70:30 mol % they are amorphous. The crystalline phasea have been identified by using ASTM data. From the naked eye observation, it is clear that viscosity of the melt decreases with the increase in Li_2O content. The possibility of crystallization to occur is more in the melt with decreased viscosity. Inspite of this, Cu-drum quenching has provided amorphous material only due to its faster quenching rate. It is evident from Table II that the heat transfer capacity of copper is more than, that of Aluminium and stainless steel. Thus, it can be concluded that, the mould material contributes towards increasing the quenching rate.

From the XRD patterns displayed in Figure 1, it can be seen that the TR samples are more crystalline than those of Al-mould. In case of Cu-drum, 75 Mol % sample got crystallised due to a considerable decrease in the content of glass former. The glass formation of 70 mol % Li_2O is further supported by DTA analysis. As Li_2O concentration increases, O/B (Oxygen to boron) ratio increases and the glass transition temperature T_g, decreases. This is tabulated in Table III.

In the light of XRD and DTA, electrical conductivity has been interpreted. Functional dependence of electrical conductivity on temperature is plotted for all compositions in Figures 2-4. The conductivity values for these samples at 483 K have been tablulated in Table IV.

It is very clear from the above figures that, the conductivity increases with the rise in Li_2O concentration, for all the compositions, except for 50:50 Li_2O:B_2O_3 samples quenched in TR and Al-mould.

From Table IV, it is clear that, 40:60 composition quenched in Al-mould gives more conductivity than TR and Cu-drum. This can be explained on the basis of DTA results which show an increase in T_g for Cu-drum as compared to Al-mould (Table III). The increase in T_g, increases the softening temperature of the system which decreases the mobility of the charge carriers, causing a decrease in conductivity. For the same compostion, conductivity has been decreased by an order of magnitude for TR sample. This is due to pelletization and sintering.

In case of 50:50 mol % conductivity has decreased for Al-mould and TR samples, due to crystalline phases present therein. As Li_2O content increases beyond 50 mol %, even crystalline sample show rise in conductivity. When the crystalline phase increase further, the conductivity decreases in comparison with Cu-drum quenched samples. Thus, the decrease in conductivity for 60:40 and 70:30 mol % of Al-mould and TR samples may be attributed to the presence of crystalline phases. The maximum conductivity has been achieved in 70:30 mol % $Li_2O:B_2O_3$ Cu-drum sample. Beyond this ratio the samples were crystallized and hence gave a reduction in the conductivity, this goes well in agreement with Tuller's report [10].

In the present invesitgation the samples giving maximum conductivity of composition $Li_2O:B_2O_3$, 70:30 quenched in Cu-drum and TR techbniques are subjected to complex impedance analysis (See Figs. 5 and 6). From these figures following facts can be understood:

1) In case of Cu-drum quenched sample at a fixed temperature, the high frequency data is represented by a nearly perfect circular arc, which passes through origin. With increasing temperature, the high frequency arc shifts to higher frequencies and its size also decreases resulting in a smaller intercept on the real axis. The spike observed at lower frequencies develops into a straight line inclined at an angle to the real axis, which is attributed to electrode polarization. The bulk resistance can be calculated by taking the intercept of an arc on the real axis, excluding electrode polarization.

2) In case of TR quenched sample (See Fig.6) two distinguishable semi-circles along with a straight line, corresponds to the bulk (armorphousness), crystalline phase and electrode polarization. The presence of semicircle in the mid-frequency range is solely due to the occurance of crystalline phase corresponding to $Li_2B_2O_4$ resulted herein because of less quenching rate in case of TR technique. When temperature is increased in this case also diameter of arcs are found to be decreased suggesting an enhancement in the conductivity.

As it is evident from the Table I that, samples beyond 50 mol % of Li_2O content got crystallized in Al-mould, and TR technique, and the crystalline phase were identified to be present in the form of traces in the glassy network of amorphous $Li_2O : B_2O_3$ system. Hence under such circumstances, it is expected that the presence of a separate crystalline phase in the glass develops interfacial boundaries separating the glass and crystalline phase which obstructs the motion of mobile cation. Also there is a presence of grain boundaries within the crystalline phase which also acts as a barrier to the cationic conduction process. In an attempt to verify above facts the crystallized samples were subjected to the impedance analysis with an expectation that, these two contributions may have two different relaxation times and hence give rise to two semicircles. The results of the impedance study are displayed in Figures 5 and 6. Impedance analysis showed similar results which agreed well with that of XRD data.

Since in case of Cu-drum, the quenching rate is higher as compared to Al-mould and TR, it was thought worth while to make a quantitative analysis of the quenching rate. Ruhl has suggested the following relation for calculating the quenching rate in case of Cu-drum [11].

$$T_F = T_d + (T_m - T_d) \exp(-Kt/c\, l^2\, d).$$

T_F = Final Flake temperature (slightly below glass transition temperature), T_d = drum temperature, T_m = melt temperature, t = contact time, K = thermal conductivity of drum material used, c = heat capacity at constant pressure of glass, 1 = thickness of glass, and d = density of glass flake. This equation involves T_g, thus for different T_g values the quenching rate will be different and so would be the conductivity. By using T_g values for different compositions, the calculated quenching rate is tabulated in Table V for the Cu-drum technique.

It is evident from this table that, as lithium content increases, the quenching rate decreases. Thus for particular quenching technique rate can be related to chemical compostion.

The present study shows that, the amorphous material gives more conductivity due to increased disorder in the system, and decrease T_g with increase in Li_2O concentration.

Thus, it can be concluded that, the quenching technique does affect the conductivity of the material.

Acknowldegement:

Authors are grateful to the Defence Research and Development Organization, Government of India, for providing financial support for this work.

Table I : Identification of crystalline phases present in Li_2O : B_2O_3 system quenched in Al-mould TR and Cu-drum

Composition :	Al-mould		TR		Cu-drum	
Li_2O : B_2O_3	Phase	d value in A°	Phase	d value in A°	Phase	d value in A°
40 : 60	Complete amorphous	-	Complete amorphous	-	Complete amorphous	-
50 : 50	$Li_2B_2O_4$	3.3510	$Li_2B_2O_4$	2.8312	Complete amorphous	-
60 : 40	B_2O_3	3.4140	$Li_2B_2O_4$	3.5466, 2.190	Complete amorphous	-
70 : 30	Unidentified phases	4.1907, 2.6217, 2.0713, 2.7143	$Li_2B_2O_4$	3.3417, 2.4232	Complete amorphous	-

Table II : Heat transfer capacity and quenching rate for different techniques.

Quenching technique	Metal	Heat transfer in cal/cm s K	Quenching rate for Li_2O : B_2O_3(40:60)
Al-mould	Aluminium	0.53	10^2K/sec
Twin-roller	Stainless steel	0.12	10^4K/sec
Cu-drum	Copper	0.94	10^7K/sec

Table III : Compositions, O/B ratio and T_g for three techniques.

Composition Li_2O:B_2O_3	O/B	T_g in K		
		Al-mould	TR	Cu-drum
40 : 60	1.83	726.3	731	735.5
50 : 50	2.0	-	-	685.7
60 : 40	2.25	-	-	598.6
70 : 30	2.66	-	-	555.7

Table IV : Comparison of conductivity at 483 K for three quenching techniques.

Composition Li_2O:B_2O_3	Conductivity (cm)$^{-1}$ at 483 K		
	Al-mould	TR	Cu-dru
40 : 60	1.55×10^{-5}	9.24×10^{-6}	1.25×10^{-5}
50 : 50	1.64×10^{-5}	1.30×10^{-6}	1.12×10^{-4}
60 : 40	5.83×10^{-5}	3.28×10^{-5}	1.43×10^{-4}
70 : 30	6.54×10^{-5}	7.34×10^{-5}	3.94×10^{-4}

Table V : Composition and quenching rate for Cu-drum.

Composition $Li_2O : B_2O_3$	Quenching rate in °K/sec
40 : 60	11.84×10^6
50 : 50	9.85×10^6
60 : 40	7.43×10^6
70 : 30	6.53×10^6

Reference:

01. K. Nassau, J. Non-cryst. Solids 42, (1980), 423-432.

02. D. Ravanie and J.L. Souquate, Phys. Chem. Glasses, 18 (1977), 27.

03. K. Nassau, R.J. Cava, S.S.I. 2, (1981), 163-170 2,.

04. H.L. Tuller and D.P. Button, Proceedings of the 6th RISQ International Symposium on Metallurgy and Material Science, (1985), 119-137.

05. Riku Ota and Naohiro Soga, Glastech, Ber. 56K Bd, 12 (1983).

06. L. Boehm and C.A. Angell, J. Phy. Chem.

07. D. P. Button, et al., S.S.I 9 & 10, (1983), 585-592, and Ota et al., Yogyo-Kyokai Shi, 1, (1984), 92.

08. C.T. Moynihan, A.J. Easteal, T. Wilder, J. Tucker, J. Phy. Chem, 78, (1974), 2673.

09. H.L. Tuller, D.P. Button and D.R. Uhlmann, J. Non-Cryst. Solids, 40, (1980) 93-118.

10. Y. Ito, K. Miyauchi and T. Oi, J. of Non-crystalline Solids 57, (1983), 93-118.

11. R.C. Ruhl, Mat Sci. Eng., 1, (1967), 313.

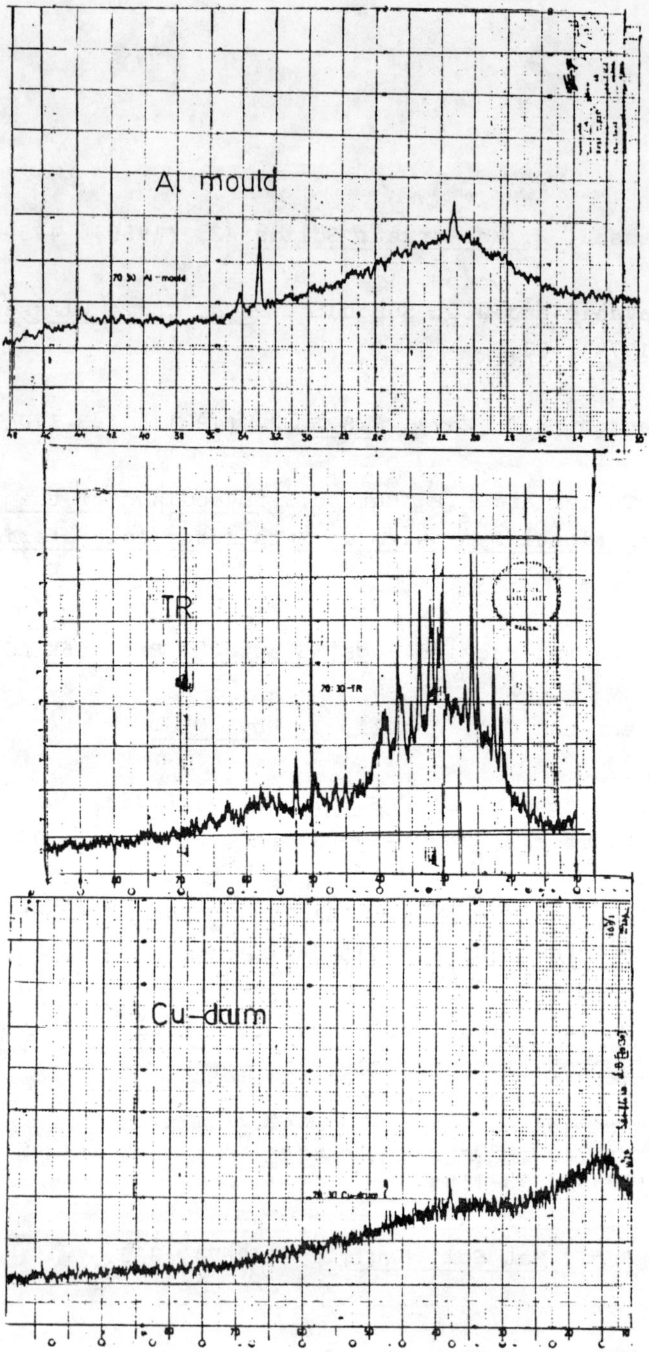

Fig.1. XRD patterns for 70:30 $Li_2O:B_2O_3$ system.

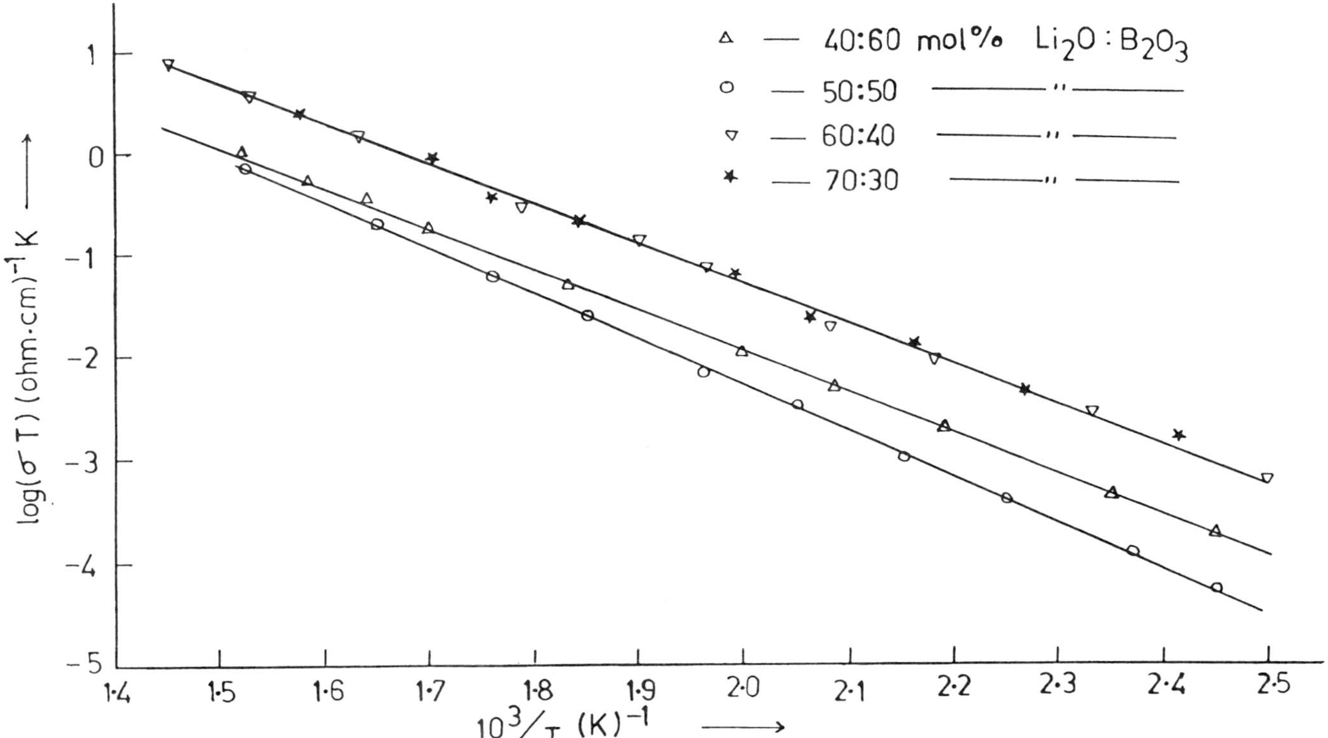

Fig. 2. Variation of electrical conductivity for $Li_2O:B_2O_3$ system with temperature for Al-mould quenched samples.

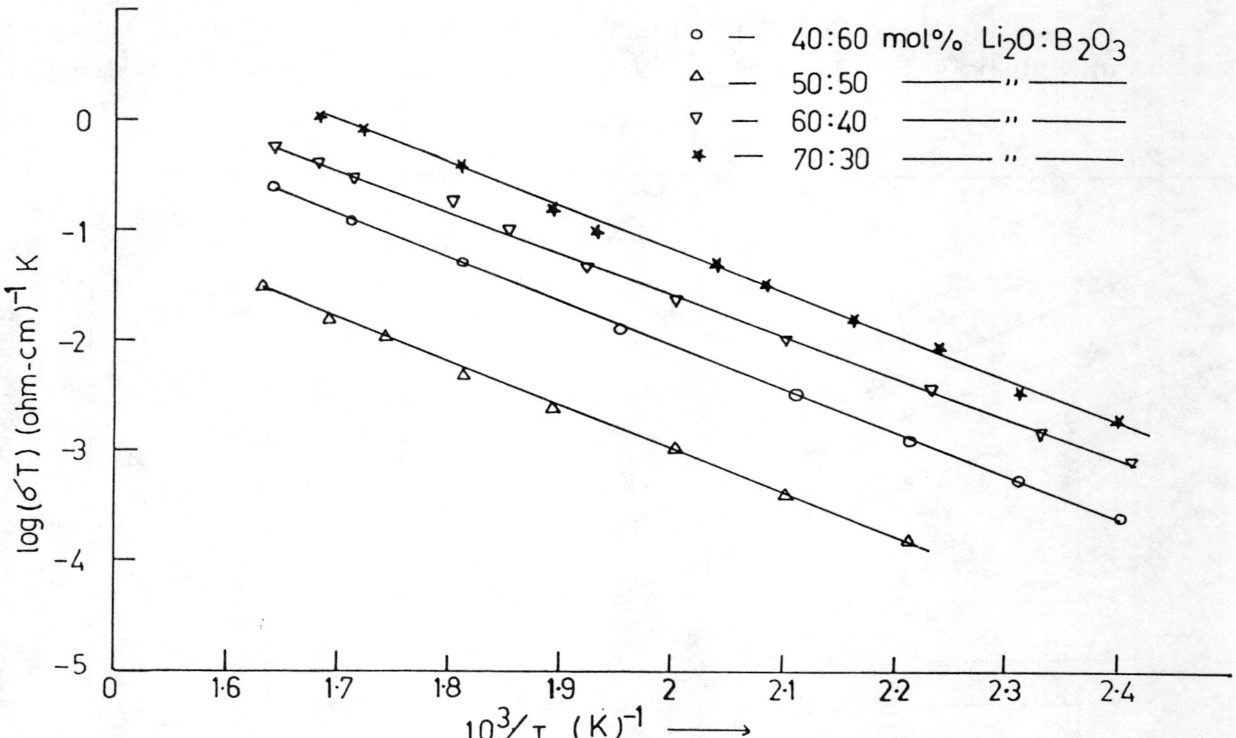

Fig. 3. Variation of electrical conductivity for $Li_2O:B_2O_3$ system with temperature for TR quenched samples.

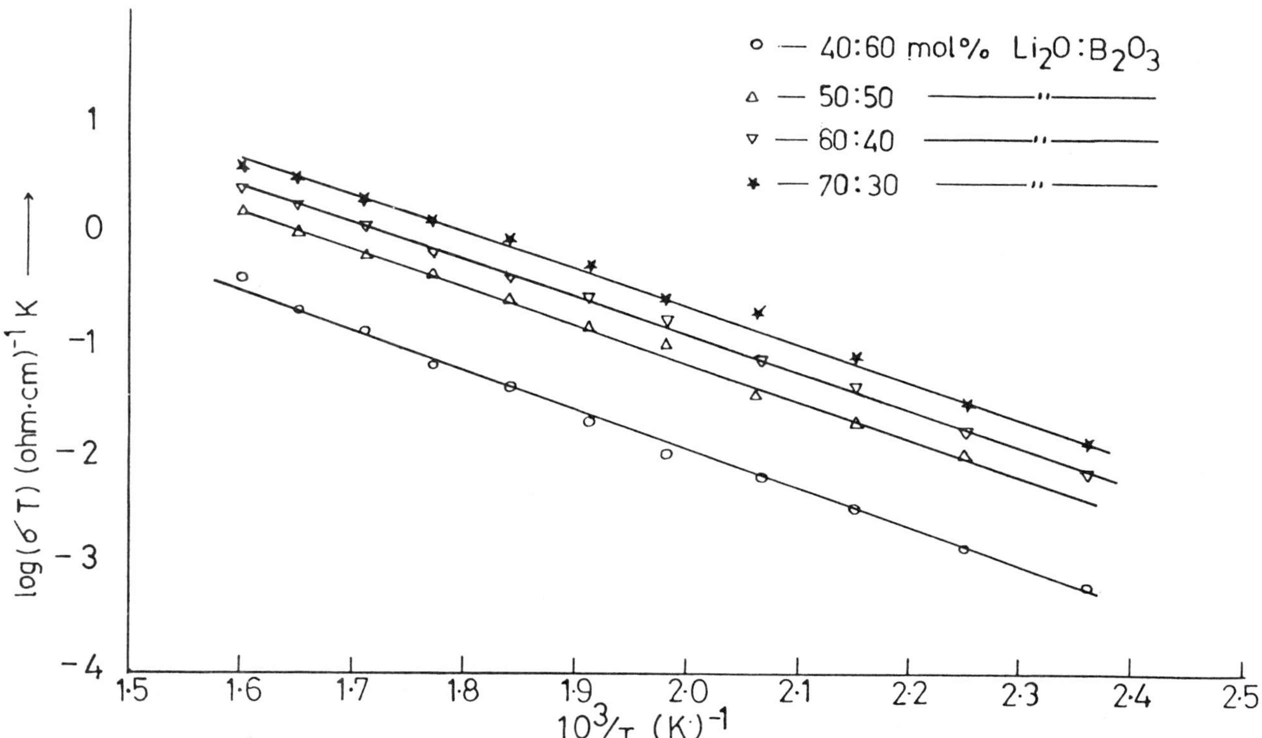

Fig. 4. Variation of electrical conductivity for $Li_2O:B_2O_3$ system with temperature for Cu-drum quenched samples.

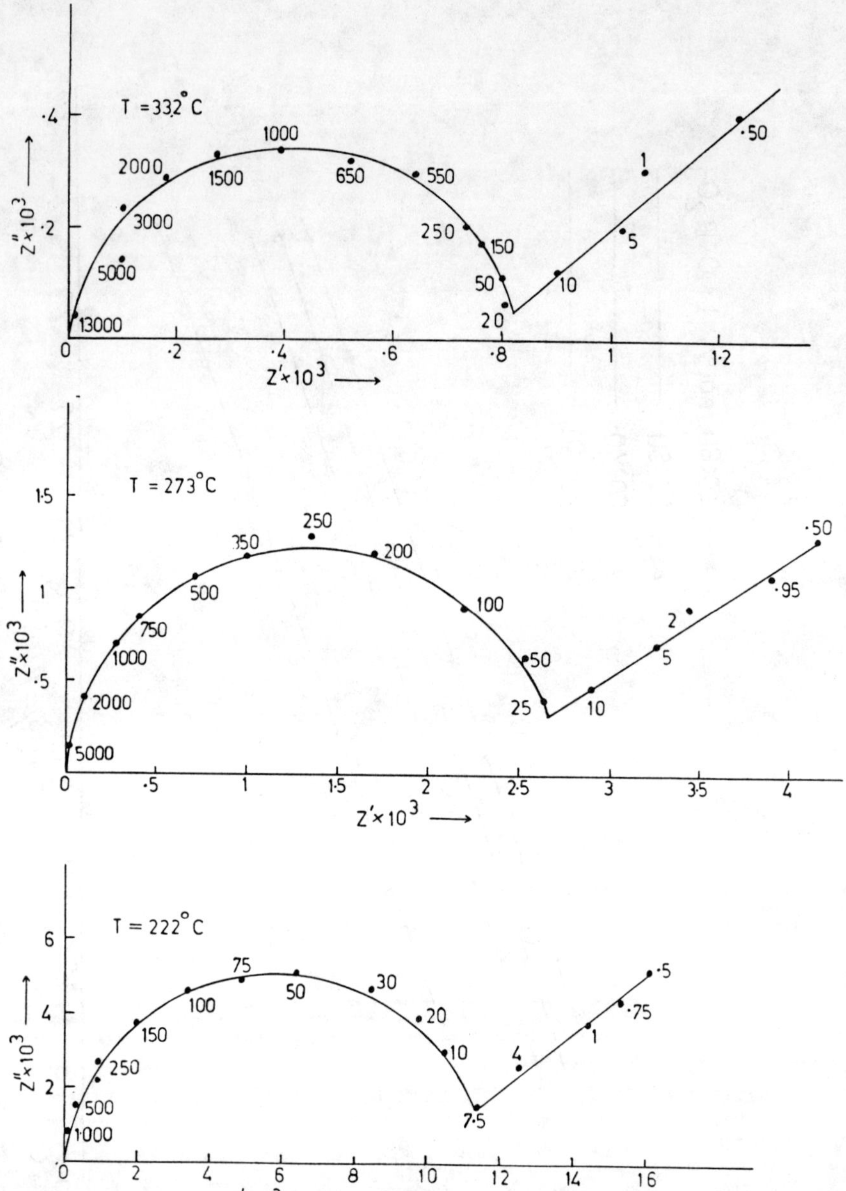

Fig. 5. Complex impedance diagram for the 70:30 Cu-drum quenched sample at different temperatures. (numbers on the plots indecates frequencies in KHz).

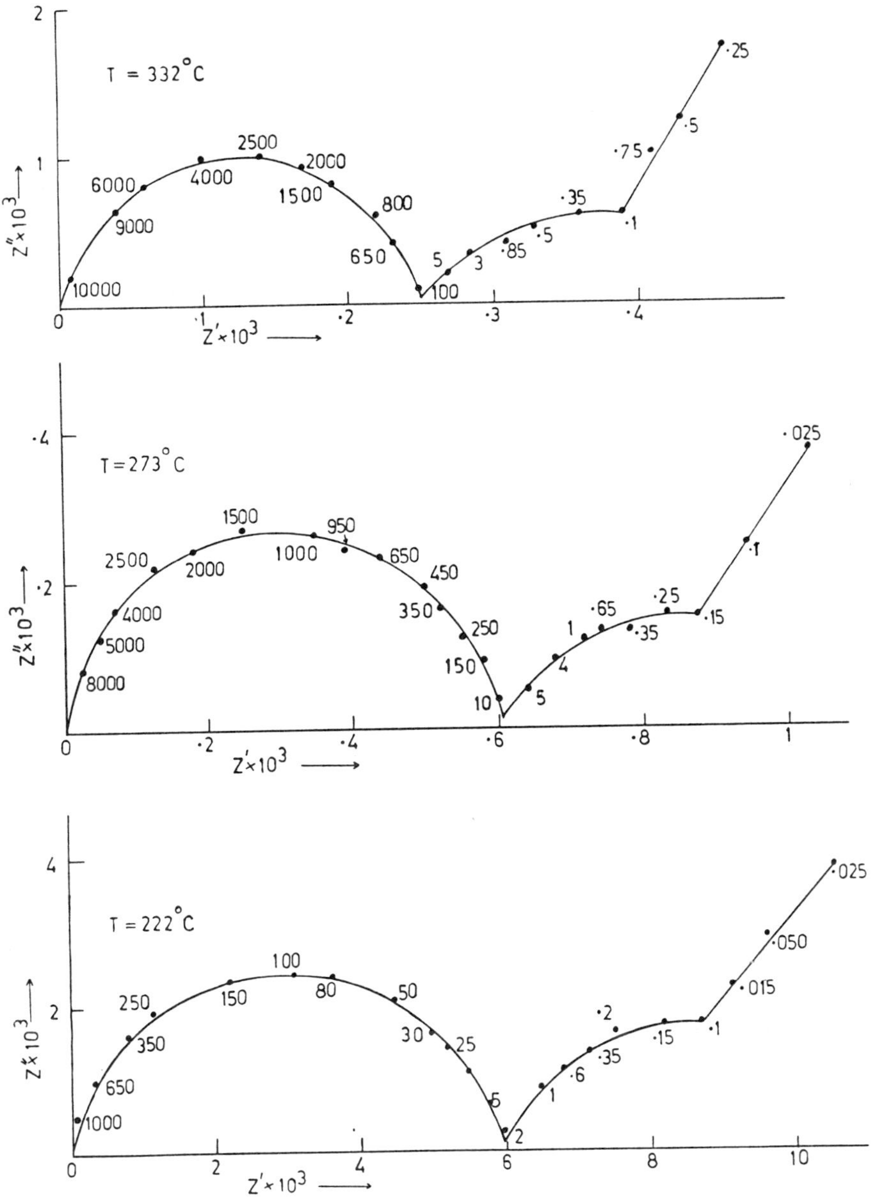

Fig. 6. Complex impedance diagram for the 70:30 TR quenched sample at different temperatures. (numbers on the plots indicates frequencies in KHz).

ANNEXURE I

PROGRAMME FOR MONDAY APRIL 20, 1987

SESSION 1	Chairman Professor Cyril Ponnamperuma
	Overview Lectures.
10:00 - 11:00	"Quantum Phenomena in Small Semiconductor Devices and Structures - I" Professor Karl-Fredrick Berggren
11:00 - 12:00	"Semiconductor Materials for Photoelectro-chemical Cells and Photocatalysts - I" Professor Kirthi Tennakone
12:00 - 01:15	LUNCH
SESSION 11	"Quantum Phenomena in Small Semiconductor Devices and Structures - II" Professor Karl-Fredrick Berggren
02:15 - 03:15	"Semiconductor Materials for Photoelctro-chemical Cells and Photocatalysts - II" Professor Kirthi Tennakone
03:15 - 03:45	TEA

SHORT PRESENTATIONS

03:45 - 04:00	"Refractive Index and Optical Gap and Annealed Amorphous Silicon" R Salam, S A Rahman, M R Muhamad and A R Danker
04:00 - 04:15	"n-Cu_2O/p-Cu S Heterojunction Thin-Film Solar Cells W Siripala and K Premasiri Kumara

PROGRAMME FOR TUESDAY APRIL 21, 1987

SESSION 111	Chairman Professor Chen Liquan
08:30 - 09:30	"Quantum Phenomena in Samall Semiconductor Devices and Structures - III" Professor Karl-Fredrick Berggren
09:30 - 10:30	"Topics in the Theory of Electron Transport for Quasi-2D and Quasi-1D Electron Gases - I" Professor P N Butcher
10:30 - 11:00	TEA
11:00 - 12:00	"Structural and Synthetic Aspects of Solids - I" Professor J Gopalakrishnan
12:00 - 01:00	"Topics in the Theory of Electron Transport for Quasi-2D and Quasi-1D Electron Gases - II" Professor P N Butcher
01:00 - 02:00	LUNCH
SESSION 1V	Chairman Dr Muhamad Yahaya
02:00 - 03:00	"Experimental Electrical Transport Properties of Low Dimensional Systems - I" Dr B L Gallagher
03:00 - 04:00	"Hot Topics in Semiconductor Physics - I" Dr Zafar Iqbal
SHORT PRESENTATIONS	
04:30 - 04:45	"Photocatalytic Nitrogen Reduction with Visible Light on Hydrous Ferric Oxide" S Punchihewa
04:45 - 05:00	"AC Conductivity of $CuInSe_2$" K Premaratne, S H S P Samrappuli, M A K L Dissanayak and S Chatraporn
05:00 - 05:15	"Band Splitting in Cubic Lattice Within the Metal Insulator Transition Model" M Ahmed

PROGRAMME FOR WEDNESDAY APRIL 22, 1987

SESSION V

Chairman
Professor Steen Skaarup

08:30 - 09:30	"Topics in Theory of Electron Transport for Quasi-2D and Quasi-1D Electron Gases - III" Professor P N Butcher
09:30 - 10:30	"Experimental Electrical Transport Properties of Low Dimensional Systems - II" Dr B L Gallagher
10:30 - 11:00	TEA
11:00 - 12:00	"Structural and Synthetic Aspects of Solids - III" Professor J Gopalakrishnan
12:00 - 01:00	"Theory of Heavily Doped Semiconductors" Professor Virul Sa-yakanit
01:00 - 02:00	LUNCH

SESSION V1

Chairman
Professor Karl-Fredrick Berggren

02:00 - 03:00	"Hot Topics in Semiconductor Physics - II" Dr Zafar Iqbal
03:00 - 04:00	TEA

SHORT PRESENTATIONS

04:00 - 04:15	"Subsolidus Phase Equlibria in the Ternary System Ca Na PO_4-Ca_2SiO_4-$Ca_3(PO_4)_2$" R P Gunawardena
04:15 - 04:30	"Dielectric Properties of Barium Borate Glass Bonded Lead Zirconate Ceramic" K Singh, and R Kahandewal
04:30 - 04:45	"Electrical Conduction in Dithiocarbamate Complexes of Mo(III), Fe(III) and Cr(III)" H M N Bandara, O A Ileperuma, P A G Dharmasena and M A K L Dissanayake
04:45 - 05:00	"Photovoltaic Properties of Semiconductor-Insulator-Semiconductor; ITO on Germanium" M Yahaya

PROGRAMME FOR FRIDAY APRIL 24, 1987

SESSION VII — Chairman: Professor J Gopalakrishnan

08:30 - 09:30	"Structural Chemistry of Framework Silicates" Professor R P Gunawardena
09:30 - 10:30	"Phase Transition of Dispersed Systems - I" Professor Chen Liquan
10:30 - 11:00	TEA
11:00 - 12:00	"Phase Transitions of Dispersed Systems - II" Professor Chen Liquan
12:00 - 01:00	Solid State Lithium Ion Conductors for Moderate Temperature" Professor Steen Skaarup
01:00 - 02:00	LUNCH

SESSION VIII — Chairman: Professor R P Gunawardena

SHORT PRESENTATIONS

02:00 - 02:15	"Phase Transition Shift of Core Level Binding energies of Some Group 1A and 11A Metals" J S H Q Perera
02:15 - 02:30	"Polarization and Discharge Characteristics of Thin Film Solid State Cells with Cu^+ Ion Conductors" M A K L Dissanayake and J Karunamuni
02:30 - 02:45	"Microcomputer Based Complex Impedence Measurements on Solid State Ionic Materials" B V R Chowdari and K Radhakrishnan
02:45 - 03:00	"Electrical Conductivity of the Li_2SO_4-$MgSO_4$ Eutetic System" M A Careem and M A K L Dissanayake
03:00 - 03:15	"Discharge Characteristics of Solid State Cells with CuI-Al_2O_3 Electrolyte" P A G Dharmasena, O A Ileperuma and M A K L Dissanayake
03:15 - 03:30	"Effect of Quenching on the Electrical Conductivity of Li_2O-B_2O_3 System" K Singh, R P Gandhi, N R Choudhary and S V Rokade
03:30 - 04:14	General Discussion
04:15 - 04:30	TEA

ANNEXURE II

LIST OF INVITED SPEAKERS

01. Professor P N Butcher
Department of Physics
University of Warwick
Coventry
England

02. Professor K F Berggren
Department of Physics and
Measurement Technology
S-58183, Linkoping
Sweden

03. Professor B L Gallagher
Department of Physics
University of Nottingham
University Park
Nottingham
England Na 7 2RD

04. Professor J Gopalakrishnan
Solid State and Structural
Chemistry Unit
Indian Institute of Science
Bangalore, 560012
India

05. Professor Z Iqbal
Department of Physics
Quaid-I-Azam University
Islamabad
Pakistan

06. Professor Chen Liquan
Institute of Physics
Chinese Academy of Sciences
Beijing
China

07. Professor V Sa-yakanit
Department of Physics
Chulalongkorn University
Bangkok 10500
Thailand

08. Professor S Skaarup
Physics Laboratory III
Technical University of Denmark
Building 309 C
DK-2800, Lyngby
Denmark

09. Dr M Careem
Department of Physics
University of Peradeniya
Peradeniya
Sri Lanka

10. Professor R P Gunawardana
Department of Chemistry
University of Peradeniya
Peradeniya
Sri Lanka

11. Professor K Tennakone
Department of Physics
University of Ruhuna
Matara
Sri Lanka

LIST OF PARTICIPANTS

01. Dr M Ahmed
    Associate Professor
    Department of Physics
    University of Dhaka
    Dhaka 2
    Bangladesh

02. Dr S Chatraphorn
    Department of Physics
    Chulalongkorn University
    Bangkok 10500
    Thailand

03. Dr R Salam
    Associate Professor
    Solid State Physics Laboratory
    Department of Physics
    University of Malaya
    59100 Kuala Lumpur
    Malaysia

04. Dr (Ms) K Singh
    Department of Physics
    Nagpur University
    Nagpur 440010
    India

05. Dr K Radhakrishnan
    Department of Physics
    Faculty of Science
    National University of Singapore
    Lower Kent Ridge Road
    Singapore 0511

06. Dr M Yahaya
    Faculty of Physical and
    Applied Sciences
    Department of Physics
    Universiti Kebangsaan Malaysia
    (National University of Malaysia)
    43600 Bangi
    Malaysia

07. Mr J S Ambalangodage
Department of Physics
University of Ruhuna
Matara

08. Mr S Wickremanayake
Department of Physics
University of Ruhuna
Matara

09. Ms D P Withana
Department of Physics
University of Ruhuna
Matara

10. Mr K P Kumara
Department of Physics
University of Kelaniya
Kelaniya

11. Dr W P Siripala
Department of Physics
University of Kelaniya
Kelaniya

12. Dr K T L De Silva
University of Colombo
Colombo 3

13. Dr H M N Bandara
Department of Chemistry
University of Peradeniya
Peradeniya

14. Dr G Dharmasena
Department of Chemistry
University of Peradeniya
Peradeniya

15. Dr O A Ileperuma
Department of Chemistry
University of Peradeniya
Peradeniya

16. Dr Q Perera
Department of Chemistry
University of Peradeniya
Peradeniya

17. Ms S U Chandrasena
Department of Physics
University of Peradeniya
Peradeniya

18. Dr M A K L Dissanayake
Department of Physics
University of Peradeniya
Peradeniya

19. Mr R R Jayaratne
Department of Physics
University of Peradeniya
Peradeniya

20. Mr B Jayasuriya
Department of Physics
University of Peradeniya
Peradeniya

21. Mr J Karunamuni
Department of Physics
University of Peradeniya
Peradeniya

22. Mr S Kodituwakku
Department of Physics
University of Peradeniya
Peradeniya

23. Mr S Premalal
Department of Physics
University of Peradeniya
Peradeniya

24. Dr K Premaratne
Department of Physics
University of Peradeniya
Peradeniya

25. Mr S Samarappuly
Department of Physics
University of Peradeniya
Peradeniya

26. Mr S Sivanantharajah
Department of Physics
University of Peradeniya
Peradeniya

27. Mr M Nizar
Department of Physics
University of Peradeniya
Peradeniya

28. Mr P Udugama
Department of Physics
University of Peradeniya
Peradeniya

29. Mr T Wickremasinghe
Department of Physics
University of Peradeniya
Peradeniya

30. Mr C N Wijayasekara
Department of Physics
University of Peradeniya
Peradeniya

31. Ms S Punchihewa
Institute of Fundamental Studies
Hantana Road
Kandy

SUBJECT INDEX